Disclaimer

The publisher of this book is by no way associated with the National Institute of Standards and Technology (NIST). The NIST did not publish this book. It was published by 50 page publications under the public domain license.

50 Page Publications.

Book Title: Certification of a Polystyrene Synthetic Polymer, SRM 2888

Book Author: Charles M. Guttman; William R. Blair; B M. Fanconi; R J. Goldschmidt; William E. Wallace III; S Wetzel; David L. VanderHart

Book Abstract: The certification of a polystyrene standard reference material, SRM 2888, is described. The M_w of SRM 2888 was determined by light scattering to be $7.19 \times 10^{+3}$ g/mol with a sample standard deviation of $0.14 \times 10^{+3}$ g/mol. A combined expanded uncertainty for this light scattering M_w determination, including systematic and random uncertainties, was estimated to be $0.57 \times 10^{+3}$ g/mol. M_n was determined by NMR analysis of the end groups and found to be $7.05 \times 10^{+3}$ g/mol with an estimated expanded uncertainty of $0.55 \times 10^{+3}$ g/mol. Fourier transform infrared spectroscopy (FTIR) and matrix-assisted laser desorption ionization time of flight mass spectrometry (MALDI-TOF-MS) were used to analyze end groups on the polymer. The polystyrene was analyzed by MALDI-TOF-MS for the number (M_n) and mass (M_w) average of the molecular mass distribution (MMD) as well as the MMD. The bottle-to-bottle variation of the SRM was checked by size exclusion chromatography (SEC) and found to be negligible. This polymer was also used in an interlaboratory comparison of MALDI-TOF-MS of synthetic polymers. The interlaboratory comparison and its results are described. Agreement between the MMD moments obtained by light scattering and NMR and the moments of the MMD obtained by MALDI in the interlaboratory comparison is found to be good overall. However, all the experimental values obtained by MALDI-TOF-MS fell below the classical values.

Citation: NIST SP - 260-152

Keyword: MALDI-TOF-MS;SRM;SRM 2888;standard reference material;synthetic polymers

NIST SPECIAL PUBLICATION 260-152

U. S. DEPARTMENT OF COMMERCE/Technology Administration
National Institute of Standards and Technology

Standard Reference Materials→

Certification of a Polystyrene Synthetic Polymer, SRM 2888

Charles M. Guttman, William R. Blair, Bruno M. Fanconi,
Robert J. Goldschmidt, William E. Wallace,
Stephanie J. Wetzel, and David L. VanderHart

National Institute of Standards and Technology
Technology Administration, U.S. Department of Commerce

NIST Special Publication 260-152
Standard Reference Materials→

Certification of a Polystyrene Synthetic Polymer, SRM 2888

Charles M. Guttman, William R. Blair, Bruno M. Fanconi, Robert J. Goldschmidt, William E. Wallace, Stephanie J. Wetzel, and David L. VanderHart

Polymers Division
Materials Science and Engineering Laboratory
National Institute of Standards and Technology
Gaithersburg, MD 20899-8540

U.S. DEPARTMENT OF COMMERCE, *Donald L. Evans, Secretary*
TECHNOLOGY ADMINISTRATION, *Phillip J. Bond, Under Secretary of Commerce for Technology*
NATIONAL INSTITUTE OF STANDARDS AND TECHNOLOGY, *Arden L. Bement, Jr., Director*

Issued November 2003

Certain commercial equipment, instruments, or materials are identified in this paper in order to specify the experimental procedure adequately. Such identification is not intended to imply recommendation or endorsement by the National Institute of Standards and Technology, nor is it intended to imply that the materials or equipment identified are necessarily the best available for the purpose.

National Institute of Standards and Technology Special Publication 260-152
Natl. Inst. Stand. Technol. Spec. Publ. 260-152, 51 pages (November 2003)
CODEN: NSPUE2

U.S. GOVERNMENT PRINTING OFFICE
WASHINGTON: 2003

For sale by the Superintendent of Documents, U.S. Government Printing Office
Internet: bookstore.gpo.gov — Phone: (202) 512-1800 — Fax: (202) 512-2250
Mail: Stop SSOP, Washington, DC 20402-0001

Certification of a Polystyrene Synthetic Polymer, SRM 2888

Charles M. Guttman, William R. Blair, Bruno M. Fanconi, Robert J. Goldschmidt,
William E. Wallace, Stephanie J. Wetzel and David L. VanderHart

Polymers Division
National Institute of Standards and Technology
Gaithersburg, MD 20899

Certain commercial materials and equipment are identified in this paper in order to specify adequately the experimental procedure. In no case does such identification imply recommendation or endorsement by the National Institute of Standards and Technology, nor does it imply necessarily the best available for the purpose.

According to ISO 31-8, the term "Molecular Mass" has been replaced by "Relative Molecular Mass," symbol M_r. Thus, if this nomenclature and notation were followed in this publication, one should write $M_{r,w}$ instead of the historically conventional M_w for the weight average Molecular Mass with similar changes for M_n, M_z, and M_v. M_w would be called the "Mass Average Relative Molecular Mass." The conventional notation, rather than the ISO notation has been used in this publication.

Table of Contents

Abstract...4
1.0 Introduction..5
2.0 Preparation, Bottling, and Sampling of SRM 2888...5
   2.1 Preparation..5
   2.2 Bottling and Sampling of SRM 2888..5
3.1 Homogeneity Testing on SRM 2888...5
     3.2 Statistical Method to Compare Chromatograms..6
        3.2.1 Match Factor..6
4.0 M_n by NMR..7
   4.1 Solution and Solvent Preparation..7
   4.2 NMR Methods...7
   4.3 Analysis of NMR Data...8
5.0 Determination of M_w of SRM 2888 by Light Scattering..9
   5.1 Light Scattering on the Polystyrene Standard...9
       5.1.1 Solution and Solvent Preparation..10
       5.1.2 Determination of dn/dc..10
       5.1.3 Refractive Indices of Solvent and Calibrant..11
       5.1.4 Light Scattering Methods...12
   5.2 Analysis of Light Scattering Data...12
   5.3 Results for SRM 2888..13
   5.4 Estimation of Uncertainties Due to Systematic Effects in the Light Scattering................14
       5.4.1 Contribution to Uncertainty Resulting from the Presence of Two Low Molecular Mass Impurities in SRM 2888..14
       5.4.2 Indices of Refraction..17
       5.4.3 Literature Value of dn/dc for Aqueous NaCl and Calibration of the Differential Refractometer...17
       5.4.4 Value of dn/dc of SRM 2888..17
       5.4.5 Wavelength of Radiation..18
       5.4.6 Rayleigh Ratio of Benzene...18
       5.4.7 Polarizer Uncertainties...18
       5.4.8 Ratio of Working Standard Scattering to Sample Scattering..........................19
       5.4.9 Solvent Density...19
       5.4.10 Solvent and Solute Masses..19
       5.4.11 Reflection Correction..19
       5.4.12 Instrumental Misalignment..20
       5.4.13 Refraction Correction...20
       5.4.14 Anisotropy of Solute...20
       5.4.15 Cutoff of Virial Expansion for SRM 2888...20
       5.4.16 Solute Degradation...21
       5.4.17 Summary...21
6.0 FTIR Determination of the t-Butyl End Groups..21
7.0 MALDI-TOF- MS Analysis of Molecular Mass Distribution.....................................22
   7.1 Introduction...22
   7.2 Preliminary MALDI-MS Analysis at NIST..23
   7.3 Interlaboratory Protocol..23

- 7.4 Estimators of the Molecular Mass Distribution (MMD)..24
- 7.5 Description Of Overall Data...24
  - 7.5.1 Mean Moments and Histogram of MMD..24
  - 7.5.2 Outliers...24
  - 7.5.3 Low and High Molecular Masses in the MMD...24
  - 7.5.4 Instrument Calibration...25
- 7.6 Effect Of Parameters On The MMD...25
  - 7.6.1 Statistical Methods to Describe the Data..26
- 7.7 Discussion of MALDI-TOF-MS..27
- 8.0 Conclusions..28
  - Acknowledgment...28
- 9.0 References..30
- Tables...33
  - Table 1: Source of Uncertainty for Light Scattering Measurements on SRM 2888................33
  - Table 2: Moments of MMD from interlaboratory Comparison using MALDI-ToF-MS.........34
- Figures...35
  - FT-IR Spectrum of SRM 2888..35
  - Mass Distribution of SRM 2888 by Mass Spectrometry..36
  - Secondary Series in Mass Spectrum of SRM 2888..37
  - Distribution of Mass of SRM 2888...38
  - Distribution of M_n Reported in Interlaboratory Comparison..39
  - Distribution of End Group Mass Reported in Interlaboratory Comparison...........................40
- Appendices..41
  - Appendix I: Participating Laboratories In Interlaboratory Comparison................................41
  - Appendix II: Uncertainty in PS Low M_w due to dn/dc Varying as a Function of Molecular Mass...42
  - Appendix III: Estimation of (dn/dc) of a Small Amount of Cyclohexane in Toluene.............44
  - Appendix IV: Certificate of Analysis, SRM 2888..45

Abstract

The certification of a polystyrene standard reference material, SRM 2888, is described. The M_w of SRM 2888 was determined by light scattering to be $7.19 \times 10^{+3}$ g/mol with a sample standard deviation of $0.14 \times 10^{+3}$ g/mol. A combined expanded uncertainty for this light scattering M_w determination, including systematic and random uncertainties, was estimated to be $0.57 \times 10^{+3}$ g/mol. M_n was determined by nuclear magnetic resonance (NMR) analysis of the end groups and found to be $6.96 \times 10^{+3}$ g/mol with an estimated expanded uncertainty of $0.40 \times 10^{+3}$ g/mol. Fourier transform infrared spectroscopy (FTIR) and matrix-assisted laser desorption ionization time of flight mass spectrometry (MALDI-TOF-MS) were used to analyze end groups on the polymer. The polystyrene was analyzed by MALDI-TOF-MS for the number (M_n) and mass (M_w) averages of the molecular mass distribution (MMD), as well as the MMD. The bottle-to-bottle variation of the SRM was examined by size exclusion chromatography (SEC) and found to be negligible.

This polymer was also used in an interlaboratory comparison of the analysis of polystyrene by MALDI-TOF-MS. The interlaboratory comparison and its results are described. Agreement between the MMD moments obtained by light scattering and NMR and the moments of the MMD obtained by MALDI-MS in the interlaboratory comparison is found to be good overall. However, all the experimental values obtained by MALDI-MS were lower than the classical values.

1.0 Introduction

This report describes certification of the M_w of SRM 2888, a polystyrene (PS), by light scattering. As supplemental numbers, the M_n by NMR end group analysis and the M_w, M_n and the MMD by MALDI-TOF-MS are also given. SRM 2888 was prepared commercially for NIST Polymers Division for use as a standard for Size Exclusion Chromatography and as a narrow distribution polymer standard with well-defined classical molecular mass moments for comparative studies by MALDI-MS. This material was specially prepared and characterized for well-defined end groups. After packaging of this material for future distribution, the M_w was determined by light scattering and M_n determined by solution NMR. End groups were further studied by FTIR and MALDI-TOF-MS. The MMD was studied by MALDI-TOF-MS. The bottle-to-bottle homogeneity of this Standard Reference Material (SRM) was examined by size exclusion chromatography (SEC).

2.0 Preparation, Bottling, and Sampling of SRM 2888

2.1 Preparation

The PS used for this SRM was prepared commercially by Polymer Source (Dorval, Québec, Canada). The polymer was specially prepared by anionic polymerization with well-defined styrene and tertiary butyl end groups. From the preparation chemistry, the polymer is expected to be an atactic polystyrene of the form:

$(CH_3)_3C[CH_2-CHPh]_n-CH_2-CH_2-Ph$ Ph = phenyl   [2.1]

The material was received in the form of a powder.

2.2 Bottling and Sampling of SRM 2888

In the following, the containers holding SRM 2888 will be referred to as vials. A total of about 450 samples of SRM 2888, about 0.3 g each, were bottled in amber vials. The entire set of samples was divided into 12 subsets. One vial was randomly selected from each subset of SRM 2888 for homogeneity testing. The first and last bottles were also taken for study. Furthermore, two bottles containing about 15 g each for use in later experiments were sampled. These are called Division Supply SRM 2888 supply in the following discussion.

3.1 Homogeneity Testing on SRM 2888

Homogeneity testing was accomplished using SEC. A Waters Alliance 2000 GPC Liquid Chromatograph (Waters Corp., Milford, MA) with a differential refractive index (DRI) detector and two Styragel 300 mm x 7.5 mm ID 10 µm HT6-E columns and one Styragel 300 mm x 7.5 mm ID 10 µm HT-2 column (Waters Corp., Milford, MA) were used in this study. The chromatography was run at 1.0 mL/min solvent flow rate. The injector and column compartment of the Alliance 2000 were controlled at 40 °C for all measurements. Tetrahydrofuran, THF, (Mallinkrodt Specialty Chemicals, Paris, KY) with added antioxidant, 2,6-di-tert-butyl-4-methyl phenol (commonly known as butylated hydroxytoluene or BHT), was used as the solvent. Toluene (Mallinkrodt Specialty Chemicals, Paris, KY) at a concentration of 5 µL/mL of THF was added to the solutions as a SEC pump marker.

Two solutions were made from each polymer sample vial. The polystyrene samples were dissolved in THF at a concentration of approximately 1.5 g/L. The order of preparing the solutions and the order of acquiring the chromatograms was randomized [1]. SEC was performed on these solutions using two injections from each solution.

After baseline subtraction, the SEC chromatograms were normalized to unit peak height and compared initially by overlaying to decide if there were visible differences outside the noise. The chromatograms from different solutions all superimposed on each other. This preliminary comparison showed that polymer samples taken from all the vials produced identical chromatograms. Statistical analysis on the chromatographic results, reported in Section 3.2, confirmed the visual observations.

3.2 Statistical Method to Compare Chromatograms

3.2.1 Match Factor

In previous SEC studies of SRMs the match factor was used to compare one chromatogram with all the others. In this study, the match factor for chromatogram i is defined as the correlation coefficient between chromatogram i and the average chromatogram of the entire testing series. Huber [2] defined the match factor as

Match Factor = $10^3 \{\Sigma x^*y - (\Sigma x^* \Sigma y)/p\}^2 / [\{\Sigma x^2 - \Sigma x^* \Sigma x/p\}\{\Sigma y^2 - \Sigma y^* \Sigma y/p\}]$. [3.1]

The value of x is the measured signal in the ith chromatogram and y is the measured signal from the average chromatogram at the same elution time; p is the number of data points in the chromatogram. The sums are taken over all data points.

At the extremes, a match factor of zero indicates no match and 1000 indicates an identical chromatogram. Generally, values above 990 indicate that the chromatograms are similar. Values between 900 and 990 indicate some similarity between chromatograms, but the result should be interpreted with care. All values below 900 are interpreted as an indication of different chromatograms [2,3,4].

The chromatograms were acquired in groups of eight solutions on different days. Match factors against the average chromatogram of all the runs were obtained for both the main SRM 2888 peak and the pump marker peak of toluene. Match factor analysis on measurements taken on one particular day revealed that both the pump marker peak and the main peak drifted a small amount during the day. This was assumed to result from a small drift in the flow rate. During any day the maximum effect was less than 0.1 cm^3 in the about 30 cm^3 of elution volume of the pump marker peak. To compensate for this, the chromatograms were shifted so that the pump marker peak for each chromatogram was always at the same elution volume. Applying this shift, all chromatograms from the main SRM 2888 peak in this study had match factors against the mean chromatogram of greater than 0.998. An ANOVA study using OMNITAB [5] made on the match factors obtained from the chromatograms indicated that the match factors of chromatograms from vials of SRM 2888 were not different using a significance level with 0.05.

The vials of SRM 2888 are indistinguishable from one another based on the above considerations.

4.0 M_n by NMR

4.1 Solution and Solvent Preparation

Two samples of SRM 2888 were weighed into separate, clean 5-mm NMR tubes. After loading the samples, the tubes were then constricted near their tops for ease of subsequent flame sealing. Then the tubes were placed on a vacuum line and pumped for about 5 min for drying, even though the SRM 2888 powder was not expected to pick up significant moisture (less than 0.0001 g change in the mass was found in a 0.3 g sample upon sitting in a balance in humid air for 20 min). Attached to this vacuum line was another vessel with degassed benzene-d6 (Mallinckrodt Specialty Chemicals, Paris, KY). Sufficient benzene-d6 was then distilled into each NMR tube so that the resulting SRM 2888 solutions were about 5 % and 13 % by mass fraction. Benzene-d6 was also distilled into an empty tube for examination of the solvent impurities. After distillation, the samples were frozen, using a liquid nitrogen bath. Then nitrogen gas was introduced into the tubes at a pressure of about 50,000 Pa (0.5 atm); this partial pressure of nitrogen helps to prevent refluxing of the solvent during measurement. The tubes were then flame sealed.

Other 5 % solutions of SRM 2888 in $CDCl_3$ were made by simply adding 0.6 mL of $CDCl_3$ to the SRM 2888 powder in a 5 mm NMR tube. These solutions were investigated by NMR mainly to ascertain the chemical identity of the 2 major impurities residing in the SRM 2888 powder. (see additional discussion in Section 5.4.1).

4.2 NMR Methods

Proton NMR spectra at 400 MHz were acquired at ambient temperature on a WM-400 spectrometer manufactured by Bruker Instruments, Inc. Resolution was found to be adequate for evaluating the integrals of interest, for both the 5 % and the 13 % solutions, under these conditions.

Spectra were taken with single pulse excitation and a pulse tip angle of 30° in order to limit the influence of slower longitudinal relaxation times on signal amplitudes. Delay between pulses was set at 20 s, confirmed to be conservatively long for giving quantitative results for all protons. Resonances of the SRM 2888 were centered in a total spectral width more than twice the range of the PS resonances in order to avoid amplitude biases from the analogue filters in the signal channel. Signal accumulations after 64 scans had adequate signal-to-noise in the Fourier transform spectra for evaluating the integrals of interest. The Fourier transforms were made large enough, by zero-filling, so that the relative integrals, for even the narrowest lines, were not limited by digital resolution.

No tetramethylsilane (TMS) was added to these samples as a chemical shift standard. It was deemed more important to have reliable integrals in the high-field region of the spectra, rather than accurate chemical shifts. For a chemical shift reference, a very weak impurity line, resonating at the highest-field position, was selected. Its shift was designated to be 0.1 ppm, presuming that it derived from a contaminant associated with vacuum greases of the polysiloxane variety. This reference gave shifts for the other resonances in reasonable agreement with expectations for PS resonance positions. The chemical shifts that will be quoted have relative uncertainties less than 0.03 ppm, but absolute uncertainties of 0.20 ppm.

4.3 Analysis of NMR Data

Integrals of the NMR spectra of the SRM 2888 in C_6D_6 were measured. The assumed structure for SRM 2888 of $(CH_3)_3$-C-$[CH_2$-CHPh$]_n$-CH_2-CH_2-Ph and the integrals (of both aromatic and aliphatic PS protons plus the end-group methyl protons) measured from the NMR, yielded an M_n of 7.05×10^3 g/mol based on measurements from the 5 % solution and an M_n of 6.95×10^3 g/mol, based on the measurements of the 13 % solution. The 13 % solution spectrum had slightly inferior resolution, as expected from a more concentrated solution, but smaller corrections, arising from protonated solvent species, had to be made to the measured aromatic integrals than for the 5 % solution. Thus, similar confidence levels were estimated for the two solutions. The above values of M_n are based on averages of the integrals of the aliphatic and aromatic proton spectral regions. As a comment on self-consistency, for each sample, there is less than a ±0.6 % variation in M_n if the aliphatic or aromatic integrals were used by themselves for calculating M_n. This provides some perspective on the integration corrections to the PS resonances discussed in the following two paragraphs. These corrections involve interfering resonances associated with both the two principal impurities in the SRM 2888 and the protonated solvent species.

The interfering resonances in the region of the aromatic protons of PS include a weak, solvent-related singlet associated with benzene-d_5-h_1 as well as the aromatic multiplets associated with the toluene impurity (see Section 5.4.1). The total intensity of the latter multiplets can be estimated from the chemical formula of toluene and the integrated intensity of the toluene methyl protons at 2.12 ppm. The intensity of the benzene-d_5-h_1 peak can be estimated from the fact that the pure benzene had two major proton peaks, one at 7.15 ppm (benzene-d5-h1) and another unidentified peak at 0.41 ppm. Since the latter peak occurred in all spectra and could be cleanly integrated, the benzene-d_5-h_1 peak could be estimated based on the ratio of the 2 peaks in the spectrum of the pure solvent. In any case, the contribution of this latter peak to the aromatic integral was less than 0.6 %.

The region of the aliphatic protons of PS overlaps two narrow impurity peaks (see Section 5.4.1) which show sharp, single resonances at 1.41 (methyl protons of toluene) and 2.12 ppm (cyclohexane). The former is situated on the upfield wing of the methylene proton resonance and the latter is found in the central resonance region of the methine protons. Because of the significant line width differences, it was possible to integrate separately the narrower impurity and broader PS lineshape contributions in spite of the overlap. Based on signal integration the total number of protons involved in these 2 impurity peaks is 0.010 times the total number of styrene protons.

The spectrum of the pure solvent gave no significant impurity peaks in the vicinities of the aliphatic PS proton resonances or the end-group methyl resonances. It is worth mentioning that M_n values were also obtained from the $CDCl_3$ solutions of SRM 2888. While these spectra contained more interfering solvent-borne impurities, including an H_2O peak that overlaps the PS aliphatic region, similar corrections to the pertinent integrals can be applied. For 3 independent samples, the M_n values were (6850, 6980, and 6950) g/mol, in excellent agreement with the M_n values in the C_6D_6 solutions. The ratio of the corrected aliphatic and aromatic integrals was always within 0.5 % of its theoretical value in these spectra; this is taken as a strong indication that the integral corrections applied were appropriate with very small systematic bias.

From the above considerations a conservative estimate of the NMR-determined M_n is $(6.96 + 0.20/-0.50) \times 10^3$ g/mol where the uncertainty represents the total expanded uncertainty. While 0.20×10^3 g/mol represents the uncertainty associated with repeatable integral measurements and absolute impurity/solvent corrections, the larger negative uncertainty of 0.50×10^3 g/mol includes a possible systematic uncertainty arising from the fact that the range of integration of the end-group methyls is more limited than that of the PS resonances. Resonances associated with the polymer, in contrast to those of the solvent molecules and low-molecular-mass impurities, tend to have wider wings. For weaker resonances, such as the end-group resonances, it is more difficult to identify the widest components compared to the widest components of a very strong line, e.g. the PS aliphatic line. By limiting the integral range, the very widest components are eliminated; hence, there may be a slight, up to 4 % in this case, underestimate of the end-group integral and a consequent overestimate of M_n.

5.0 Determination of M_w of SRM 2888 by Light Scattering

5.1 Light Scattering on the Polystyrene Standard

5.1.1 Solution and Solvent Preparation

The polystyrene samples were weighed on an analytical balance with 0.01 mg resolution in mass indication. Buoyancy corrections were applied.

Samples of SRM 2888 were not subjected to special drying before or during weighing since PS powder is not generally considered to take up moisture. Less than 0.0001 g change in the mass was found in a 0.3 g sample after exposure to humid air for 20 min.

The response of the balance was tested by weighing a 50 mg standard balance mass. The balance appeared to arrive at equilibrium mass within 1 min after the 50 mg standard mass had been loaded onto the pan, and maintained the same indicated apparent mass within ± 0.00001 g random variation.

Analytical reagent grade toluene (Mallinckrodt Specialty Chemicals, Paris, KY) was used as the solvent without modification. The amount of solvent added was determined gravimetrically on a top loading balance with 0.01 g resolution.

The SRM 2888 sample was placed in solvent the afternoon before the light scattering measurements were made. The solution bottles were quiescent over night, then stirred with magnetic stirrers for about 30 min the next morning. The partial specific volume for PS in toluene tabulated as 0.917 mL/g by Brandrup and Immergut [6], and toluene density versus temperature tabulated from several sources by Riddick and Bunger [7], were applied to compute solution concentrations at 25.0 °C.

Solvent and solution samples were filtered into light scattering cells. The solvent and solution samples were filtered through filtering assemblies with double thickness of Millipore Fluoropore membrane (Millipore Corp., Bedford, MA) with 0.22 μm average pore size.

5.1.2 Determination of dn/dc

The differential refractive index for SRM 2888 in toluene at 25 °C for light of 632.8 nm wavelength was determined using a LDC/Milton Roy Chromatix KMX-16 (Thermo-Separation Products, Sunnyvale, CA) differential refractometer. The differential refractometer was calibrated against aqueous NaCl solutions.

Refractive increments versus concentration for several aqueous electrolyte solutions at several wavelengths of light were reported by Kruis [8]. Refractive increments for the same solutions at the He-Ne laser wavelength, λ = 632.8 nm, have been determined from interpolation of the data in the Kruis tables [9,10]. A cubic equation for these refractive increments as a function of NaCl concentration in aqueous solution at 25 °C is given in the manual for the differential refractometer. This equation was used to compute the refractive increments of the aqueous NaCl solutions prepared as standards in calibrating the differential refractometer.

Mallinckrodt analytical reagent NaCl was dried in a vacuum oven at 90 °C for three days in preparation to be used as a calibrant. The dried NaCl was then maintained in a vacuum desiccator except while taking salt samples to prepare solutions. Distilled water was degassed by boiling and left to cool to ambient temperature overnight in storage bottles tightly capped with zero headspace. The storage bottles had been leached out with several changes of boiling distilled water before being used to contain the degassed distilled water. Both salt and water components of each solution were measured gravimetrically, and atmospheric buoyancy corrections were applied to compute the concentrations as g NaCl/100 g H_2O. Measurements in the differential refractometer were conducted on seven solutions ranging in concentration from 0.5 g NaCl/100 g H_2O to 2.0 g NaCl/100g H_2O in intervals of 0.25 g NaCl/100 g H_2O. The calculated refractive increments of the solutions were fitted to their average image displacements, dn/dx, to generate a linear calibration equation of refractive increment versus image displacement.

Refractive increments between solvent and solutions of SRM 2888 in toluene were determined on solutions that had been prepared on the morning of each day during which the measurements were conducted. The solutions were prepared by the procedure described in Section 5.1.1.

Image displacement measurements were conducted on four toluene solutions of SRM 2888 varying in concentration from approximately 3 g/L to 24 g/L at 25 °C. An average was taken from eight individual image displacement measurements for each solvent versus solvent and solution versus solvent pair. The average image displacement determined for each solution was bracketed by the determination of average solvent versus solvent image displacements before and after that of the solution. The incremental image displacement by each solution was obtained by subtracting the mean of the bracketing solvent average image displacements from the average image displacement by the solution. The refractive increment of each solution was computed by application of the calibration equation to the incremental image displacement of the solution. Linear regression analysis of the refractive increments versus the concentrations of the solutions yielded a differential refractive index, dn/dc = 0.103 mL/g for SRM 2888 in toluene with a standard deviation of the mean of 0.0009 mL/ g.

The differential refractive index at 632.8 nm for other comparable polystyrenes in toluene had also been determined earlier by others with values ranging from 0.081 mL/g to 0.108 mL/g [11,12]. Applying the formula of Hadjichristdis and Fetters [12] yielded an estimate of 0.105 mL/g for the dn/dc of SRM 2888. The value of 0.108 mL/g obtained for SRM 2888 is considered acceptable given the fact that end group differs from the polystyrene reported [11].

5.1.3 Refractive Indices of Solvent and Calibrant

The refractive indices of toluene and of benzene, the latter used as the Rayleigh scattering standard, were derived from tabulated values for the He-Ne laser wavelength, 632.8 nm, at 23 °C by Kaye and McDaniel [13].

5.1.4 Light Scattering Methods

Light scattering measurements on the SRM 2888 solutions in toluene were made on a Brookhaven Instrument Model BI-200 (Brookhaven Instrument Corp., Ronkonkoma, NY) light scattering apparatus equipped with a 10 mw He-Ne laser light source. The V_v polarization is selected for the scattered light intensity since the laser beam is vertically polarized and a vertical polarizer is used in the detector optics.

The temperature was controlled at 23.0 °C in all experiments with SRM 2888 in toluene. In all experiments, the intensity measuring system was calibrated with the intensity of the light scattered from the beam at 90° angle by a benzene standard cell.

5.2 Analysis of Light Scattering Data

Light scattering data at V_v polarization from polymer solutions of concentration c and scattering angle Θ may be analyzed by fitting the scattering signal $I(\Theta,c)$ to [14]

$$I(\Theta,c) = I(\Theta,0) + c\, I_G / \{(\sin \Theta) \sum_{ij} c^i \sin^{2j}(\Theta/2)\} \quad [5.1]$$

In eq. [5.1], I_G is the scattering signal from the benzene working standard at $\Theta = 90°$.

We must first decide how many terms on the right-hand side must be included to provide an adequate fit to the experimental data. The dependence of c/I_c, where $I_c = \sin\Theta[I(\Theta,c)-I(\Theta,0)]/I_G$, upon c and upon $\sin^2(\Theta/2)$ reflects solute-solvent interactions and solute size, respectively. At the low molecular mass of SRM 2888 no measurable size dependence was expected. Preliminary measurements over angles from 37.5° to 142.5° confirmed this. Thus, angle-averaged values of the Zimm function (the left hand side of eq [5.2]) were used for analysis at any one concentration. Also angles from 60° to 120° were taken on all further measurements.

c/I_c versus a power series in c up to c^2 was looked at to see if it would provide an adequate fit. The analysis revealed that the linear approximation was adequate at concentrations below 20 g/L for SRM 2888 (see below).

Thus the analysis assumed the normal Zimm form

$$c\, I_G / [\sin \Theta \{I(\Theta,c) - I(\Theta,0)\}] = (C_{00} + C_{10}c + C_{20}c^2) \quad [5.2]$$

The coefficients in eq. [5.2] are related to the M_w, mean-square radius of gyration of the polymer, R_G^2, and the second virial coefficients, A_2, by [14]:

$$M_w = (K'C_{00})^{-1} \quad [5.3]$$

$$A_2 = 2\, K'C_{10} \quad [5.4]$$

$$A_3 = K' C_{20}/3 \quad [5.5]$$

$$K' = 4\pi^2 n_B^2 (dn/dc)^2 / (\lambda_0^4 N_A V_v^B) \quad [5.6]$$

where:

λ_0 is the wavelength in vacuum of the scattered light, 632.8 nm in this work, n and n_B are the indices of refraction of the solvent and benzene taken as 1.494 and 1.497, respectively calculated as described in 5.1.3, dn/dc is the differential refractive index of the solution, measured as described in 5.1.2, N_A is Avogadro's number, taken as 6.022×10^{23} /mol, V_v^B is the Rayleigh ratio for the vertically polarized scattering of vertically polarized light from benzene, used for calibration and obtained as described in the following paragraph.

The "vertical-vertical" Rayleigh ratio V_v is related to the Rayleigh ratio R_v for the unpolarized scattering of vertically polarized He-Ne laser and the depolarization ratio ρ_v for polarized light by:

$$V_v^B = R_v^B/(1+\rho_v) \quad [5.7]$$

Using the published [13] values for benzene

$$R_v^B = 12.6_3 \times 10^{-6} \text{ cm}^{-1}$$

and $\rho_v = 0.265$, yields

$$V_v^B = 9.98 \times 10^{-6} \text{ cm}^{-1}$$

5.3 Results for SRM 2888

Six sets of light scattering solutions were made from SRM 2888 using toluene as solvent. Each set consisted of four to five independently made up solutions. The light scattering on each solution set was run twice, usually on consecutive days. The polymer for each solution within each set was taken from a sample vial or from a container designated Division Supply SRM 2888, Intensities initially were measured at ten scattering angles in the range from 37.5° to 142.5°. These measurements showed that there was no angle dependences arising from a radius of gyration; subsequent intensities were measured at ten scattering angles in the range from 60.0° to 120.0°. The angle averaged scattering intensities combined data from all the runs from concentrations in the concentration range from 0.0025 g/mL to 0.040 g/mL.

A preliminary fit by least squares was made to eq. [5.2] and the results were used to calculate M_w, A_2, and A_3 using eq. [5.3]-[5.7]. Light scattering gave a M_w of 7.11×10^3 g/mol with a standard uncertainty of 0.06×10^3 g/mol, an A_2 value of 0.0008 mol mL/g^2 with a standard uncertainty of 0.0001 mol mL/g^2 and an A_3 value of 0.006 mol mL/g^2 with a standard uncertainty of 0.002 mol mL/g^2.

Next the data were fit over a limited concentration range dropping the A_3 term. Using a bootstrap analysis with the A_2 and A_3 from the estimations made above, it was shown that the effects of nonlinearity in concentration on the computed M_w started at .02 g/mL. Thus, all data for concentrations below 0.02 g/mL were fit to an equation linear in concentration which yielded a M_w of 7.19×10^3 g/mol with a standard uncertainty of 0.060×10^3 g/mol, and an A_2 value of 0.00096 mol mL/g^2 with a standard uncertainty of 0.00005 mol mL/g^2. This linear fit is taken to best represent the data, and this value of M_w is chosen as the certified value. Both this fit and the fit including the A_3 term are well within the estimated uncertainty of each fit so from the point of view of the data the two numbers are indistinguishable.

The value of A_2 from this fit is in good agreement with that given by the formula obtained by Fetters et al. [15] in their review of thermodynamic properties of polymer solutions. Using M_w of 7.19×10^3 g/mol, an A_2 value of 0.00103 mol mL/g^2 is derived from the Fetters et al. formula.

In compliance with the NIST policy [16] on reporting uncertainties in measurement, the standard uncertainty reported above is multiplied by a coverage factor of 2 to obtain the component of expanded uncertainty listed in Table 1.

5.4 Estimation of Uncertainties Due to Systematic Effects in the Light Scattering

The likeliest sources of systematic uncertainty in the determination of the M_w of SRM 2888 by light scattering were described in the preceding sections. Upper limits for their magnitudes were estimated using a scheme similar to that used in reference [14] for the estimation of systematic uncertainties in SRMs 1482, 1483 and 1484. These uncertainties are listed in Table 1 for SRM 2888.

5.4.1 Contribution to Uncertainty Resulting from the Presence of Two Low Molecular Mass Impurities in SRM 2888

Solution-state proton NMR indicated the presence of two low molecular mass impurities in the SRM 2888 material. These impurities are also discussed in Section 4.3. The mass fractions of the impurities, derived from measured proton fractions that are normalized to the total amount of polystyrene protons, are obtained as follows: a) The proton NMR spectrum of impurity 1 in a 5 % SRM 2888 solution in CDCl$_3$ at ambient temperature consists of a singlet resonance at a chemical shift of 1.43 ppm with respect to tetramethylsilane (TMS); in C$_6$D$_6$, this impurity appears at 1.41 ppm. This resonance, in C$_6$D$_6$, is situated near the upfield edge of the broader PS CH$_2$ resonance; hence, the integral of the impurity can be separated reasonably well from that of the PS. The impurity is identified as cyclohexane from its singlet character and its chemical shifts in both solvents. Cyclohexane is difficult to remove by heating at temperatures below the glass transition of SRM 2888. The relative amount of cyclohexane can be estimated by evaluating its integral along with the total PS integral. The fact that cyclohexane has a proton/carbon ratio of 2:1, relative to 1:1 for PS allowed determination of relative mass

fractions from the relative proton intensities. Considering measurement uncertainties as well as variations in cyclohexane content from sample to sample yielded a maximum cyclohexane mass fraction in the samples studied of 0.22 % by mass fraction. Owing to uncertainties in establishing the proper separation between cyclohexane and PS NMR resonances, the cyclohexane integration has an uncertainty of 15 % of its value.
b) Impurity 2 also exhibits singlet resonances at a chemical shift of 2.34 ppm in $CDCl_3$ and at 2.12 ppm in C_6D_6. These singlets almost certainly arise from the methyl protons of toluene. In principle, this assignment could be verified by the aromatic spectral fingerprint of toluene, which has several bands in the 7.15 ppm to 7.30 ppm range. However, that is not so easy because of the high multiplicity and subsequent weakness of these lines plus the fact that there is strong overlap with the aromatic PS resonance and the resonances from the residual protonated-solvent species. The assignment was based, finally, on 2 observations. First, toluene was intentionally added in two small increments to the $CDCl_3$ solution and observed that the added intensity, for the methyl protons, appeared at the very same shift that was present before. Even the weak aromatic features were enhanced in a way consistent with the impurity being toluene. Second, the 2.34 ppm shift in $CDCl_3$ and the 2.12 ppm peak in C_6D_6 agree with published tables for the methyl protons of toluene in each solvent. Based on this chemical assignment and integrations of the methyl peak as well as the total PS intensity, the maximum value of the mass fraction of toluene was 0.42 % by mass.

The estimated uncertainty in M_w rising from the presence of the two low molecular mass impurities was determined from the relationship derived from eq. [5.2], [5.3] and [5.6], namely,

$$M_w = \text{Limit as c approaches zero of } [K^*/\{c_p(dn/dc)_p^2 \}] \quad [5.8]$$

In the above equation, terms in (dn/dc) and c in eq. [5.2],[5.3] and [5.6] were replaced with terms referring directly to the polymer, c_p and $(dn/dc)_p$ since it is the uncertainty caused by the replacement of c_p, the actual polymer concentration, with c_m, the measured concentration of polymer, which is object of the discussion here.

The only terms affected by the presence of low mass impurities are the change in concentration of polymer, c_p, and the change in refractive index with concentration of the polymer, $(dn/dc)_p$.

Thus

$$\delta M_w / M_w = - \delta c_p/c_p - 2\, \delta(dn/dc)_p/(dn/dc)_p \quad [5.9]$$

The contribution from each of these was estimated and added to the overall corrections in the appropriate sections

The contribution to the uncertainty from the term $\delta c_p/c_p$ in the above equation may be estimated as follows: the effect of mass is simply proportional i.e. a 1 % uncertainty in the mass of the polymer due to an impurity leads to a 1 % uncertainty in the M_w. Thus this uncertainty leads to a 0.67 % uncertainty in mass. This contribution will be considered in Section 5.4.10 where the uncertainty in the solute mass determination is discussed.

In the remaining discussion in this section, the uncertainties contributing to the term in $\delta(dn/dc)_p$ are considered in some detail.

The measured change in the refractive index in a dilute solution of concentration c_m, δn_m, may be written as

$$\delta n_m = c_m (dn/dc)_m \qquad [5.10]$$

For the dilute solutions considered here, δn_m may be expressed as the sum of contributions from each component, the polymer (δn_p), impurity 1 (δn_1), and impurity 2 (δn_2). Thus,

$$\delta n_m = \delta n_p + \delta n_1 + \delta n_2 \qquad [5.11]$$

and

$$c_m (dn/dc)_m = c_p (dn/dc)_p + c_1 (dn/dc)_1 + c_2 (dn/dc)_2 \qquad [5.12]$$

where
  c_p = true concentration of polymer,
  $(dn/dc)_p$ = change of refractive index with polymer concentration in toluene,
  c_1 = concentration of impurity 1, cyclohexane,
  $(dn/dc)_1$ = change of refractive index with concentration of impurity 1 in toluene,
  c_2 = concentration of impurity 2, toluene, and
  $(dn/dc)_2$ = change of refractive index with concentration of impurity 2 in toluene.

Then dividing by c_p in eq [5.12] and letting $y_i = c_i/c_p$ yields,

$$(1 + y_1 + y_2)(dn/dc)_m = (dn/dc)_p + y_1 (dn/dc)_1 + y_2 (dn/dc)_2 \qquad [5.13]$$

Rearranging the above equation gives,

$$(dn/dc)_p - (dn/dc)_m = y_1 \{(dn/dc)_m - (dn/dc)_1\} + y_2 \{(dn/dc)_m - (dn/dc)_2\} \qquad [5.14]$$

where $\delta (dn/dc) = (dn/dc)_p - (dn/dc)_m$

Since impurity 2 is toluene then $(dn/dc)_2 = 0.0$. From the NMR $y_2 = 0.0042$ so the contribution from the term + $y_2 \{(dn/dc)_m - (dn/dc)_2\}$ is 0.103 * 0.0042 or 0.00044, where for this work a value of 0.103 is taken for $(dn/dc)_m$

To evaluate the term $y_1 \{(dn/dc)_m - (dn/dc)_1\}$ an estimate of -0.09 (see Appendix ii) is used for $(dn/dc)_1$ of impurity 1, cyclohexane, versus toluene. Thus the term $\{(dn/dc)_m -$

$(dn/dc)_1\}$ is 0.193 and multiplying by the mass fraction $y_1 = 0.0022$ obtained from the NMR gives 0.00043 for this term.

Adding together both contributions to the $\delta(dn/dc)$ uncertainty, the total fractional uncertainty in dn/dc is 0.85 % from the impurities. The fractional uncertainty in M_w is 1.7 % or 0.13 g/mol. Since all the above are estimates of the maximum uncertainties in each component, this uncertainty is taken to be the expanded uncertainty. This contribution is given in Table 1.

5.4.2 Indices of Refraction

Following reference [14], an estimate of 0.1 % is a proper upper limit for systematic uncertainties in M_w arising from uncertainties in the literature values of solvent index of refraction.

5.4.3 Literature Value of dn/dc for Aqueous NaCl and Calibration of the Differential Refractometer

Calibration of the differential refractometer required interpolation of the data of reference [10] to the 632.8 nm wavelength used for the light-scattering measurements. An estimate for the uncertainty in the interpolated values of dn/dc is 0.6 %, due primarily to uncertainties in the interpolation process. The calibration factor determined for the differential refractometer had a relative standard deviation (rsd) of 0.097 %. Combining the above uncertainties with an allowance for possible linear uncertainties in the refractometer, gives an estimate of 1 % for the proper upper limit for uncertainty to dn/dc from this contribution. Considered as a 95 percent confidence interval estimate, this quoted uncertainty provides an expanded uncertainty in dn/dc of 1 % [16] which would contribute 2 % or 0.14×10^3 g/mol expanded uncertainty in M_w determination.

5.4.4 Value of dn/dc of SRM 2888

5.4.4.1 Uncertainty Arising From Measured Value

The differential refractive index dn/dc of SRM 2888 in toluene at a temperature of 23 °C. was determined as described in 5.1.2. The mean value obtained for dn/dc was 0.103 mL/g, with a standard deviation of the mean of 0.0009 mL/g or 0.87 %. As the dn/dc appears raised to the second power in the M_w calculation, the uncertainty in the calculated M_w resulting from the standard deviation in the mean of dn/dc is estimated as twice the estimated standard uncertainty in the dn/dc, or 1.7 %. The expanded uncertainty from this measurement is 3.4 % or 0.24×10^3 g/mole.

5.4.4.2 Uncertainty Arising from dn/dc Variation with Molecular Mass

The differential refractive index dn/dc of low molecular mass polymers shows a molecular mass dependence arising from the effects of end groups. Generally this variation is shown to be in the form

$$dn/dc = A + B/M \quad [5.15]$$

Details of the contribution due to this effect are given in Appendix ii. For this polymer from the classical measurement an estimate $M_n - M_w = -0.230 \times 10^3$ g/mol is obtained and choosing A from Hadjichristidis and Fetters as 0.108 mL/g yields the correction on the order of 0.023×10^3 g/mol. The contribution to the overall expanded uncertainty from this effect is negligible.

5.4.5 Wavelength of Radiation

For the He-Ne laser employed in this work, uncertainties in the wavelength of the radiation are completely negligible compared with uncertainties from other sources.

5.4.6 Rayleigh Ratio of Benzene

For benzene at 632.8 nm, reference [13] gives: $R_{V,V+H} = 12.6_3 \times 10^{-6}$ cm^{-1} and $\rho_v = 0.265$, giving $R_{V,V} = R_{V,V+H}/(1 + \rho_v) = 9.98 \times 10^{-6}$ cm^{-1}. In the following discussion $R_{V,V+H}$ and ρ_v are abbreviated by R and ρ, respectively. The R-values reported in reference [13] are <u>accurate</u> to 2 % (systematic) and the relative standard uncertainty for the R-value of benzene is given as 0.21/12.63, or 1.7 %. No estimates of either accuracy or precision for values of ρ were given in reference [13]. However, ρ is obtained as the ratio of two intensities, the larger of which is, or is close to, the intensity measured for the determination of R. The photomultiplier detectors were apparently operated in the current mode, and it seems reasonable to suppose that the <u>absolute</u> uncertainty in the smaller intensity is the same as that of the larger, and that the <u>relative</u> uncertainty in the larger is the same as that in R. Then if r is the relative standard deviation (rsd) of R, the standard deviation in ρ becomes sd(ρ) = r%(1 + ρ^2) and rsd(1+ρ) = [r/(1+ρ)]%(1 + ρ^2). Combining this with the rsd in R results in rsd($R_{V,V}$) = [r/(1+ρ)]%2(1 + ρ + ρ^2) , which is about 1.3r for $\rho = 0.265$. The product, 1.3r = 1.3x1.7 % yields a standard uncertainty 2.2 %. This standard uncertainty combined (by root-sum-squares) with the stated standard uncertainty of 2 % for the R-values [13] yields a standard uncertainty of 3 % or 0.21×10^3 g/mol. Applying the coverage factor of 2 to this standard uncertainty results in an expanded uncertainty of 0.42×10^3 g/mol.

5.4.7 Polarizer Uncertainties

There are four polarizers to consider: First, the "vertically polarized" laser beam actually contains "horizontally polarized" components for two reasons: First, the polarizer inside the laser head lets through a small fraction ε of the "wrong" polarization; Second, the principal axis of polarization of the light from the laser may not be exactly perpendicular to the plane of the incident and scattered beams. Both will cause light assumed to be vertically polarized to contain a small admixture of horizontally polarized light. The effect upon scattering signals from SRM 2888 will be slight, but the effect upon the benzene calibration signals is to change the effective Rayleigh ratio that should have been used from the R_{VV} value toward the $R_{V,V+H}$ value. The resulting uncertainty in M_w

is $\rho\varepsilon/(1-\varepsilon)$ for the first effect and $\rho\tan^2\alpha$, where α is the angular missetting, for the second. The uncertainty from both effects together is $\rho[\varepsilon/(1-\varepsilon) + \tan^2\alpha]$.

Second, in an exactly analogous way, the analyzing polarizer in front of the detector may be nonideal and/or mispositioned. In this case, let δ be the contribution from the nonideality of the polarizer, and let ß be the angle of missetting. The resulting expression for the uncertainty is then: $\rho[\delta/(1-\delta) + \tan^2 ß]$.

Finally, since all these uncertainties are of the same sign they are added to get: $\rho[\varepsilon/(1-\varepsilon) + \tan^2\alpha + \delta/(1-\delta) + \tan^2 ß]$. From the manufacturers' specifications ε and δ are estimated to be 1/500. The uncertainty in missetting the angle of the laser is estimated as less than $\alpha = 5°$. The alignment of the polarization analyzer is assumed as better than ß = 3°. Using $\rho = 0.265$ the uncertainty is 0.265[.0020 + .0077 + .0020 + .0027] = 0.0038 = 0.4 % with at least a 95 % level of confidence considering the liberal boundaries assigned to the constituent uncertainties and their combination by linear summation instead of root-sum-of-squares. This quoted uncertainty provides an estimated expanded uncertainty of 0.4 % or 0.03×10^3 g/mol.

5.4.8 Ratio of Working Standard Scattering to Sample Scattering

Since photon counting techniques were employed, there should be no <u>systematic</u> uncertainties from this source. Random uncertainties are reflected in the overall random uncertainty of the M_w.

5.4.9 Solvent Density

The density of toluene at 25 °C, 0.86231 g/mL, given in reference [7], is estimated to be accurate to 0.1 %, or better. The resulting expanded uncertainty on M_w is just 0.1 % or less than 0.01×10^3 g/mol.

5.4.10 Solvent and Solute Masses

Solvent masses in the determinations for SRM 2888 were chosen so that the <u>solute's</u> mass was always about 0.05 g. Using the uncertainty limit of 0.1 mg normally assigned to the balance used to weigh the SRM 2888 samples yields uncertainties in the solute masses of 0.2 % (and negligible uncertainties in solvent masses). The resulting expanded uncertainty in M_w is about 0.2 % or less than 0.02×10^3 g/mol. This must be added to the contribution to the uncertainty of M_w arising from the small molecule impurities as discussed in Section 5.4.1. This contribution is about 0.67 % or 0.053×10^3 g/mol. The two contributions together give by a sum of squares 0.060×10^3 g/mol.

5.4.11 Reflection Correction

The refractive index of toluene at 23 °C and 632.8 nm wavelength is given [13] as 1.4940. The refractive index of the sample cell is given by the vendor as 1.474 at the 589 nm wavelength. Although the temperature of the toluene, and the wavelength for the refractive index of the cell, in this case are not correct for our experiments, these values should be adequate to estimate what will turn out to be an extremely small uncertainty. Substitution of these two refractive indices into Fresnel's equation for reflection from an interface between two transparent dielectrics [17] yields a reflectance factor $f = 2 \times 10^{-4}$. Comparison calculations of M_w with and without this correction for SRM 2888 show the resulting uncertainty to be less than 0.01 % or less than 0.01×10^3 g/mol.

5.4.12 Instrumental Misalignment

For the geometry of the Brookhaven light scattering instrument, it is expected that any deviation from constancy of $I \sin(\Theta)$ is indicative of instrument misalignment. The $I \sin(\Theta)$ measurements show a maximum uncertainty of 1.0 % in the intensity compared to the intensity at 90 degrees as a function of angle. The maximum misalignment estimated from this is at most $1.0°$. Examination of four data sets, expanding the range of (0.002 g/mol to 0.02) g/mol, showed that a systematic angle variation of $\pm 1.0°$ produced variations in M_w of less than 0.1 %. Since the uncertainty in alignment seems more random than consistent, the correction is expected to be less. Thus, the expanded uncertainty arising from instrument misalignment is estimated to be less than 0.1 % or 0.01×10^3 g/mol for SRM 2888.

5.4.13 Refraction Correction

A detailed analysis of the optical geometry of the light scattering instrument employed in this work can not be carried out, since the main detector optics unit was inaccessible. However, rough analyses based on assumptions about the internal geometry of the detector unit lead to an uncertainty of about 0.3 %. A reasonable uncertainty <u>limit</u> might then be about twice this, or 0.6 %, which would give an expanded uncertainty of 0.6 % or 0.04×10^3 g/mol on M_w due to refraction uncertainty.

5.4.14 Anisotropy of Solute

We know of no reported optical anisotropy of polystyrenes in toluene.

5.4.15 Cutoff of Virial Expansion for SRM 2888

As described in Section 5.2, the data fit included both linear and quadratic functions of concentration. The solution concentration range of the measurements was from (0.002 to 0.04) g/mL. The coefficients and their uncertainties of the fits for the entire range and for limited ranges are given in Table 1. Solution concentrations used for the final analyses were limited to the region where linear terms in c appeared to suffice. However, as is seen from the analyses in Table 1 fitting with linear or quadratic terms had little effect on M_w. If the linear fit is used over the entire range of the data, then the fit does not account for the curvature in the data. An estimate of the uncertainty in the calculated M_w caused by not including A_3 in the data fit is given by the change in M_w when the linear fit is made over the extended concentration range. The uncertainty in estimating M_w is about 100 g/mol; the expanded uncertainty is twice this value, or 200 g/mol.

Another estimation of the expanded uncertainty is made by fitting the data over concentrations from (0.002 to 0.02) g/mol with and without the including of the quadratic terms. These two fits give M_w that differ by no more than 100 g/mole, well within our predicted expanded uncertainty.

Finally, an angle-average Zimm function was developed assuming M_w of 7.19×10^3 g/mol and estimating A_2 and A_3 from the quadratic fit described in Section 4.3. A_2 is estimated to be 0.0008 mol mL/g^2 and A_3 is estimated to be 0.006 mol mL/g^2 (Notice these values are in very good agreement with the Yamakawa estimation of the $A_3 = k_f (A_2)^2 M_w$ where k_f is about 1.0). Using this function and concentration data at the points of our experimental points, we fit a linear functional form from (0.002 to 0.02) g/mol and found an M_w that differed from the true M_w of about 70 g/mol, much less than the 200 g/mol in our estimate of the expanded uncertainty arising from this contribution.

5.4.16 Solute Degradation

By their nature, light scattering experiments are of short duration. A number of solutions can be prepared and examined by light scattering in a single day. Solutions were made up one day and often run the next day. The second run on the same solutions often occurred the following day or a few days later. Allowing the solutions to sit around for many days had no noticeable effect. Furthermore, the SEC studies on SRM 2888 solutions produced no indication of degradation of the polymer over periods of weeks.

As long fresh solutions were prepared and examined within a day or two no solute degradation is apparent. As this was the procedure followed in the entire series of light scattering experiments there is no uncertainty arising from degradation.

5.4.17 Summary

The standard deviation of the mean of the determined M_w values, from analysis of variance of the experimental data, and the systematic uncertainties obtained from Section 5.4.1 through 5.4.16 are listed in Table 1 for SRM 2888.

The combined expanded uncertainties of SRM 2888 are computed as root-sum-of-squares of the component expanded uncertainties following the formal NIST policy for evaluating and expressing uncertainty in measurements [16]. The combined expanded uncertainty of SRM 2888 is 0.57×10^3 g/mol.

6.0 FTIR Determination of the t-Butyl End Groups

Infrared spectroscopic analysis was used to confirm the identity of the end groups and to identify the existence of minor chemical impurities that may be present at detectible levels in the as-received material. According to the synthesis, each polystyrene molecule should contain a t-butyl group at one end and a hydrogen atom at the other end (as seen in 4.1). Whereas the latter group is difficult to discern in the infrared spectrum, the t-butyl group is easily identified at the 1:70 molar ratio with styrene repeat unit expected with this polystyrene. The FT-IR spectrum of the SRM 2888, average of 200 scans at 1.0 cm^{-1} resolution, is shown as the top trace in Figure 1.

As a model infrared spectrum of polystyrene terminated by a t-butyl group, the infrared spectrum of neopentyl benzene, 2,2'-dimethylpropyl benzene, was recorded and shown as the bottom trace in Figure 1. Two bands, at 1365 cm^{-1} and 1393 cm^{-1}, are present in this spectrum that are identified with motions of the methyl groups of the t-butyl group [18]. Another intense band, also attributable to the t-butyl group, occurs at 1475 cm^{-1}. All three of these bands are evident in the infrared spectrum of the SRM 2888, top trace in Figure 1. To enhance visualization of contributions to the infrared spectrum from end groups of SRM 2888, or chemical impurities to the extent that they exist at concentrations comparable to end groups, the spectrum of high molecular mass polystyrene, SRM 1479 (M_w = 1,050,000 u), was recorded and subtracted from the spectrum of the sample polystyrene to remove the 'normal' polystyrene contributions. The resultant difference spectrum appears as the middle trace in Figure 1. The infrared difference spectrum between the SRM 2888 and SRM 1479 contains three bands at 1365 cm^{-1}, 1393 cm^{-1} and 1475 cm^{-1} that are characteristic of the t-butyl group. Although the difference spectrum contains several other bands of comparable magnitude these appear at frequencies identical to normal infrared bands of polystyrene, and for this reason these bands cannot be unambiguously assigned to end groups, or impurities. The absence of other bands in the difference spectrum at comparable or greater intensities suggests no chemical impurities are present with concentrations greater than 1 %.

7.0 MALDI-TOF- MS Analysis of Molecular Mass Distribution

7.1 Introduction

Matrix-assisted laser desorption/ionization time-of-flight mass spectrometry (MALDI-TOF-MS) [19-22] is a new and important technique in characterization of synthetic polymer molecular mass distribution (MMD)[23-30]. In this method, the sample is prepared by mixing the analyte with a UV absorbing matrix material, organic acids are common, and a cationizing salt, such as silver trifluoroacetate, in a common solvent and depositing the mixture on a sampling plate. A pulsed nitrogen laser ablates the sample mixture into the gas phase producing cationized species of the analyte for traditional time-of-flight mass spectrometry analysis. Much is still unknown about the repeatability and accuracy of the molecular mass distribution as measured by the MALDI-TOF-MS instruments. As the number of polymer analyses by MALDI-MS has increased, scrutiny of the MALDI-MS results in comparison to classically-obtained values for M_w and M_n has shown they do not always agree [22,23,25,31]. For this reason, the MMD of SRM 2888 determined by MALDI-MS is presented here as supplemental data, rather than certified results. To assess the repeatability of the method, the NIST Polymers Division initiated an interlaboratory comparison of SRM 2888 by MALDI-MS. This section describes the MALDI-MS analysis results of twenty-three respondents (laboratories) that participated in the interlaboratory comparison. A fuller discussion, including participating laboratories, is given in reference [32]. In this reference, SRM 2888 is referred to as polystyrene of nominal molecular mass 7000 u.

The outline of this section is as follows. In Section 7.2 a discussion of a preliminary MALDI-MS of the polymer is given as well as references to further research on the MALDI-MS of the polymer. In Section 7.3, a brief description of the protocol for the interlaboratory comparison is given. In Section 7.4, two kinds of descriptors of the data are defined, including traditional polymer moments of M_n, M_w and M_z. The statistical analysis of all the data using these descriptors is given in Section 7.5. The effects of various parameters are described in Section 7.6, including the distinction between and within laboratory and the effects of choice of matrix materials. Finally, in Section 7.7 summarizes some conclusions from the interlaboratory comparison.

7.2 Preliminary MALDI-MS Analysis at NIST

A preliminary MALDI analysis on SRM 2888 was conducted at NIST on a Bruker REFLEX II MALDI-TOF-MS (Billerica, MA) to ascertain that the polystyrene was consistent with the structure shown in [2.1]. The spectral main peaks, Figure 2, from a calibrated instrument agreed well with the structure in [2.1]. However, the MALDI mass spectrum revealed an unexpected secondary series of peaks, also with 104 u mass separations, in addition to the expected main series ions; see Figure 3. One possibility was that some of these intermediate peaks indicated end groups not seen in the FTIR. However, additional experimentation on the polystyrene sample revealed that the secondary series peak position changed with respect to the main series peaks when different matrices were used. Post-source decay [33] was used to determine that the secondary peaks arose from two sources: either adducts of the matrix and/or cations with the polymer or fragmentation of the polymer along the main chain. The matrix salt adducts caused the secondary peaks to shift when different matrices were used. None of the secondary peaks were attributable to additional end groups. Details of how these attributions were established are given in reference [33].

7.3 Interlaboratory Protocol

Each participating laboratory was instructed to perform MALDI-MS using two protocols involving different sample preparations. The first protocol required retinoic acid for the matrix and AgTFA for the salt [28]. The second protocol allowed each laboratory to use a sample preparation of their choosing. Each laboratory was asked to produce three MALDI spectra for each protocol to check for intralaboratory variability. Six spectra are obtained from two sample preparations for each laboratory. Each laboratory reported M_n and M_w for each spectrum as well as the integrated mass intensity signal for each separate peak of the PS mass spectrum with the cation mass subtracted from the peak masses. The M_n and M_w values used in the following discussions were obtained from the analysis of reported integrated signal peak intensities, rather than reported values of M_n and M_w owing to discrepancies in the methods used by some of the participants to determine values of M_n and M_w.

7.4 Estimators of the Molecular Mass Distribution (MMD)

To facilitate the analysis of the data, particularly in the tails of the mass distribution where issues of cut-off and baseline may influence the determined moments of the distribution, the integrated peak intensities of each data set were separated into eleven mass divisions, or bins, before comparison. The bins are taken to be six PS repeat units, 625 u, in width, except for bins 1 and 11, which contain the remaining area of the tails. The bin area of the distribution is then used for the statistical comparison.

7.5 Description Of Overall Data

7.5.1 Mean Moments and Histogram of MMD

The mean number-average molecular mass (M_n) of the entire data set, using all instruments and both protocols, was found to be 6609.89 u. The standard deviation (σ) was found to be 120.64 u, and the standard uncertainty of the mean ($\sigma/\sqrt{N}$) was 11.77 u [36] where N is 105. The standard deviation is approximately equivalent to one repeat unit of polystyrene, 104 u, giving a very narrow distribution of data. The mean of the moments M_n, M_w, and M_z, the standard deviations and the standard uncertainty of the mean are given in Table 2.

These data are also represented in Figure 4, which illustrates the mean distribution of the bins for the data obtained using protocol 1. The bins are normalized, therefore indicating the fraction of the total MMD that they contain. The bin means show that bins 1 and 2 make up less than 4 % of the MMD. Also bins 9, 10 and 11 make up less than 4 % of the MMD. The histogram of the mean bins in Figure 4 is Gaussian.

7.5.2 Outliers

A statistical analysis of M_n values was used to identify data sets that contributed to erroneous evaluations (outliers). These data sets were deleted in subsequent analysis to prevent erroneous influences. Since the M_n distribution is normally distributed, a normal distribution can be used to identify the outliers of the distribution. Three standard deviations of a normal distribution contain 99.8 % of the data; any values that lie outside of this range are considered outliers [35]. Figure 5 shows the distribution of the M_n data and the fitted normal curve for these data. Three laboratories reported mass distributions, which yielded moments that fell outside of three standard deviations of the mean of the normal distribution. For the purposes of this analysis, these data points were classified as outliers and excluded from further data analyses.

7.5.3 Low and High Molecular Masses in the MMD

Due to the method of assigning bins, some laboratories' results contained no data in the bins representing the tail regions of the MMD. For the low mass tail of the distribution 55 % of the laboratories reported no data in bin 1, and 10 % reported no data in bin 2. The missing data were more extreme in the high mass tail, presumably due to a loss of instrument sensitivity in the high mass region. In bin 10, 45 % of the laboratories' resulting data sets contained no data, and 85 % of data sets contained no data in bin 11. The loss of data in the tail regions may be a result of instrument sensitivity and resolution, or may be a result of baseline correction and integration. Each of the different instrument types produced data sets that lacked data in the tails of the distribution. Based on examination of the above-mentioned outlier data, it appears that some missing data are attributable to the influence of the integration methodology. Depending on the software used, peaks in the tail regions with baseline noise are very easily missed by peak selection software.

7.5.4 Instrument Calibration

The accuracy of the instrument calibration of each laboratory was assessed by calculation of the end group mass. The masses of the end groups of SRM 2888 were calculated by taking the difference between the mass of the maximum signal of the distribution and the calculated mass from the number of repeat units; the cation mass has already been subtracted. Figure 6 shows the distribution of the calculated end group masses. The end groups are tert-butyl and hydrogen (as seen in [2.1]) and have a total mass of 58.14 u. The calibration of most TOF mass spectrometers are expected to be accurate to less than 3 u, but as can be seen in Figure 6, some laboratories' calibrations were off over 40 u.

Consideration was given to whether the inaccuracy of the instrument calibrations would cause uncertainties in the data analysis. When compensations were made to M_n due to these calibration discrepancies, the value of the mean M_n was only slightly altered, and the variance of the M_n values decreased slightly. When the corrected M_n was analyzed, the results of analysis presented in the following sections were not altered. Therefore, the corrections were not continued in the statistical analysis described below.

7.6 Effect Of Parameters On The MMD

In the analysis of the interlaboratory comparison data, several parameters were considered as possible influences on the SRM 2888 molecular mass distribution. The parameters examined were reporting laboratory, sample preparation method, instrument manufacturer, and TOF-MS instrument mode (reflectron or linear). Whether the laboratory in which the polymer is examined has an influence on the MMD is an important test of the consistency of the MALDI-TOF-MS method of polymer characterization. The type of matrix used in sample preparation of the polymer for MALDI analysis may also be a significant parameter. The two matrices preparations compared in this analysis were all-trans retinoic acid and dithranol. Other matrix preparations were used, but not by a sufficient number of laboratories for statistical comparison. The parameter "instrument" classifies by instrument manufacture, not the model of the instrument. The mode of the instrument, linear or reflectron, was not evaluated owing to insufficient data.

7.6.1 Statistical Methods to Describe the Data

Analysis of variance (ANOVA), a standard statistical analysis tool, was used to make inferences about the data populations [35]. Both one-way and two-way ANOVA tests were used with the latter assessing effects of two parameters on the response variable. The two-way ANOVA considers that effects due to one parameter may mask the effects due to the second parameter. The effects of each factor are called main effects and one, both, or neither may turn out to be significant. In addition to these main effects (and independent of them) there may be an effect due to their interaction that accounts for how simultaneous changes in the two parameters affect the response variable.

7.6.1.1 Effect of Laboratory on the MMD

The statistical analysis of the reporting laboratory included only data that were taken using protocol 1 for sample preparation. Since each participating laboratory prepared samples a positive laboratory effect may be due also to a sample preparation effect. After outliers were identified and removed, as well as laboratories that did not include three mass spectra repeats of the polystyrene, 16 laboratories were included in the analysis.

A one-way ANOVA of the moments for the laboratory parameter showed a significant effect on the molecular mass distribution. The ANOVA of the bins for the laboratory parameter revealed a significant effect on each of the bins of the molecular mass distribution. Surprisingly, even the center bins, which are expected not to be as sensitive to the moments as the bins representing the tails, showed a large variation among laboratories.

But the one-way ANOVA of the laboratory parameter does not give conclusive results, because the instrument parameter and laboratory parameter are confounded. The confounding exists because each laboratory used only one instrument type. Therefore, two-way ANOVA was used to differentiate the two effects.

The two-way ANOVA first accounts for the effect of instrument and then the effect of laboratory. Because of the confounding of the instrument and laboratory parameters, the data set was reduced to include only those instruments run by multiple laboratories, leaving 13 laboratories in the statistical analysis. In the two-way ANOVA, when the instrument parameter is accounted for, the laboratory parameter is found to have a significant effect on all of the moments and all of the bins representing the molecular mass distribution of polystyrene.

7.6.1.2 Effect of Instrument on the MMD
The instrument variable considers all instruments from the same manufacturer together as one parameter, regardless of the model of the instrument. The six different

instrument types represented in our study were Bruker, PerSeptive, Physical Electronics' Trift, Micromass, ThermoBioanalysis Vision, and several homemade instruments. Of these, only three instrument types were used by more than one laboratory; only the 13 laboratories that ran one of these three instrument types were included in the analysis. As well, only data reported from the defined sample preparation protocol were considered in the statistical analysis.

In order to determine the effect of instrument on the molecular mass distribution, the influence of the laboratory parameter must be removed. This was achieved by taking the mean of the three determinations of moments (or bins) from each laboratory. These laboratory means can then be analyzed by a one-way ANOVA for the instrument parameter.

The ANOVA of the mean laboratory moments for the instrument parameter yielded no significant effect of instrument on the molecular mass distribution. The variation within instrument type was not significantly less than the variation among instruments. When the bins were examined by this method, only bin 8 was significantly influenced by the instrument parameter. Bin 8 represents 7 % of the MMD and represents the high mass tail of the distribution. Overall, the instrument has little influence on the molecular mass distribution.

7.6.1.3 Effect of Different Matrices on the MMD

For analysis of sample preparation, only the 6 laboratories that ran both dithranol and retinoic acid as matrices, were considered in the analysis. A two-way ANOVA of the data accounted first for laboratory effects, and then assessed the influence of the matrix on the moments and bins of the molecular mass distribution. The instrument parameter is also considered in this statistical method, because the instrument parameter is accounted by the laboratory parameter.

The two-way ANOVA results revealed that the matrix used in the sample preparation did not significantly influence the moments of the molecular mass distribution. When the bins were analyzed, only bin 3 was significantly influenced by the matrix parameter. Bin 3 includes data in the low mass tail of the polymer molecular mass distribution that represents 6 % of the MMD. Thus the matrix type may have an effect on the low mass tail of the polymer distribution.

7.7 Discussion of MALDI-TOF-MS

The M_w and M_n obtained by MALDI-TOF-MS for the interlaboratory comparison were found to be 6.74×10^3 g/mol and 6.61×10^3 g/mol, respectively, with a standard deviation of 0.11×10^3 g/mol for M_w and 0.12×10^3 g/mol for M_n. The MALDI-MS gives lower M_n and M_w than determined by classical methods. The estimated expanded uncertainties of the classical methods encompass the M_n and M_w averages obtained by MALDI-MS However, the estimated expanded uncertainty of the classical methods includes systematic uncertainty evaluated as type B uncertainty. The MALDI-MS data does not have an estimated systematic uncertainty. In fact, much of the research directed at MALDI-TOF-MS is aimed at estimating and lowering the systematic uncertainty, or type B uncertainty. However, it is noteworthy that the M_n and M_w values obtained by MALDI-MS in every reporting laboratory were lower than the M_n and M_w obtained by classical methods. The largest value of M_n reported was still 200 u less than the M_n obtained from NMR. The disagreement of the M_n and M_w obtained by MALDI-MS and the classical methods may be too great to be attributed to instrumental and statistical uncertainties alone, particularly since the bulk of the uncertainty in NMR and light scattering arise from very different causes. This combined with the fact that the standard statistical uncertainty from the MALDI-MS is so small, leads to the possibility that the systematic uncertainties in MALDI-TOF-MS may be biased in one direction. Discussion of possible sources of systematic uncertainties is given in reference [32].

8.0 Conclusions

The M_w of SRM 2888, the same material used in a MALDI-TOF-MS Interlaboratory Comparision, was certified by light scattering to have a value of M_w of 7.19×10^3 g/mol with a sample standard deviation of 0.14×10^3 g/mol and an estimated expanded uncertainty of 0.57×10^3 g/mol. The M_n, measured by end group analysis by NMR, was found to be 6.96×10^3 g/mol with an estimated expanded uncertainty of 0.40×10^3 g/mol. This is reported here as a supplemental number. The uncertainty estimates for the certified value of M_w by light scattering included both repeatability, a type A evaluation of uncertainty, and systematic uncertainties, a type B evaluation of uncertainty [17]. The M_w and M_n obtained by MALDI-TOF-MS for the interlaboratory comparison were found to be 6.74×10^3 g/mol and 6.61×10^3 g/mol, respectively, with a standard deviation of 0.11×10^3 g/mol for M_w and 0.12×10^3 g/mol for M_n. These are also reported here as supplemental values, as well as the MMD reported in Figure 4.

NMR and FTIR analysis confirmed that SRM 2888 has only one pair of end groups, as expected from the method of polymer synthesis. This is consistent with MALDI-MS results obtained at NIST. Bottle-to-bottle variability on the SRM 2888 vials was found to be below detectable levels by size exclusion chromatography.

For the MALDI-TOF-MS Interlaboratory Comparison, the ANOVA analysis of the data showed that the variation among participating laboratories was significant. The type of instrument used in obtaining the MMD had little influence on the data. Only the bin data for one bin showed an influence of instrument type. The matrices that were used in the sample preparation of the polystyrene for MALDI-MS-TOF analysis did not have a significant influence on the molecular mass distribution.

Acknowledgment

The authors thank participants from laboratories that contributed to the interlaboratory comparison. S.J.W. thanks the National Science Foundation for the Chemometric Fellowship, Prof. Nancy Flournoy of American University, and NIST for the research capabilities. C.M.G. and S.J.W. thank Dr. Mark Levenson of the NIST Statistical Engineering Division for assistance with the statistical analysis of the MALDI-MS data.

9.0 References

[1] M.G. Natrella, "Experimental Statistics", National Bureau of Standards Handbook 91, U.S. Department of Commerce, 1963.

[2] L. Huber, "Application of the Diode-Array Detector in HPLC", Hewlett Packard, France, 08/89, pg. 89-100

[3] B.J. Bauer, B. Dickens and W.R. Blair, "Chromatographic Examination of Intaglio Inks, Resins and Varnishes", NISTIR 4949, 1991

[4] C.M. Guttman, W.R. Blair, J.R. Maurey, "Recertification of SRM 1482a, a Polyethylene", NISTIR 6054, 1997, available from National Technical Information Service, Technology Administration, U.S. Department of Commerce, Springfield, VA 22161

[5] "PC-OMNITAB: An Interactive System for Statistical and Numerical Data Analysis," available from National Technical Information Service, Technology Administration, U.S. Department of Commerce, Springfield, VA 22161

[6] J. Brandrup and E.H. Immergut, editors, "Polymer Handbook, Third Edition", John Wiley and Sons, New York, 1989, section VIII pg. 93

[7] J. A. Riddick and W. B. Bunger, "Organic Solvents, Physical Properties and Methods of Purification," Wiley-Interscience, New York, 1970, Table 58, pg 145. or Table 30, pg 107.

[8] A. Kruis, Z. Physik. Chem. B $\underline{34}$, 13-50 (1936).

[9] R. J. Anderson, Appl. Opt. $\underline{8}$, 1508-1509 (1969).

[10] LDC Analytical, "Instruction Manual for KMX-16 Laser Differential Refractometer" Section 5.3, "Calculation of Calibration Constant," Riviera Beach, FL, USA

[11] K. Matsuo, W.H. Stockmayer, F. Bangerter, Macromolecules $\underline{18}$, 1346-1348 (1985)

[12] N. Hadjichristidis, L.J. Fetters, Jour. Poly. Sci. Poly Phys. Ed $\underline{20}$, 2163-2166 (1982)

[13] W. Kaye and J. B. McDaniel, Appl. Opt. $\underline{13}$, 1934-1937 (1974)

[14] Charles C. Han, Herman L. Wagner, and Peter H. Verdier, "The Characterization of Linear Polyethylene SRM's 1482, 1483, and 1484 III Weight-Average Molecular Masses by Light Scattering." NBS Special Publication 260-61, P.H. Verdier and H.L. Wagner ed. December 1978 or National Bureau of Standards, Journal of Research $\underline{83}$, 5-193 (1978)

[15] L.J. Fetters, N. Hadjichristidis, J.S. Linder, and J.W. Mays, Journal of Physical and Chemical Reference Data 23, 619-640(1994)

[16] B. N. Taylor and C. E. Kuyatt, "Guidelines for Evaluating and Expressing the Uncertainty of NIST Measurement Results." NIST Technical Note 1297, January 1993, available from National Technical Information Service, Technology Administration, U.S. Department of Commerce, Springfield, VA 22161

[17] Brookhaven Instrument Corp., "Instruction Manual for Laser Light Scattering Goniometers," May 1984, Third Printing

[18] A.S. Wexler, Spectrochim. Acta 21, 1725-1742 (1965).

[19] M. Karas, F. Hillenkamp, Anal. Chem. 60, 2299-2301 (1988)

[20] K. Tanaka, H. Waki, Y. Ido, S. Akita, Y. Yoshido, T. Yoshido, Rapid Commun. Mass Spectrom. 2, 151-153 (1988)

[21] F. Hillenkamp, M. Karas, R.C. Beavis, B.T. Chait, Anal. Chem. 63, 1193 (1991)

[22] U. Bahr, A. Deppe, M. Karas, F. Hillenkamp, U. Giessman, Anal. Chem. 64, 2866 (1992)

[23] G. Montaudo, M.S. Montaudo, C. Puglisi, F. Samperi, Rapid Commun. Mass Spectrom. 9, 453-460 (1995)

[24] G. Montaudo, D. Garozzo, M.S. Montaudo, C.Puglisi, F.Samperi, Macromolecules 28, 7983-7989 (1995)

[25] P.O Danis, D. E. Karr, W.J. Simonsick, D.T. Wu, Macromolecules 28, 1229-1232 (1995)

[26] G. Montaudo, M.S. Montaudo, C. Puglisi, F. Samperi, Rapid Commun. Mass Spectrom. 9, 1158-1163 (1995)

[27] G. Montaudo, E. Scamporrino, D. Vitalini, P. Mineo, Rapid Commun. Mass Spectrom. 10, 1551-1559 (1996)

[28] D.C. Schriemer, L. Li, Anal. Chem. 68, 2721-2725 (1996)

[29] M.W.F. Nielen, S. Malucha, Rapid Commun. Mass Spectrom. 11, 1194-1204 (1997)

[30] H. Zhu, T. Yalcin, L. Li, J. Am. Soc. Mass Spectrom. 9, 275-2819 (1997).

[31] H.M. Burger, H.M. Muller, D. Seebach, K.O. Burnsen, M. Schar, H.M. Widmer, Macromolecules 26, 4783 (1993).

[32] C.M. Guttman, S.J. Wetzel, W.R. Blair, B.M. Fanconi, J.E. Girard, R.J. Goldschmidt, W.E. Wallace, and D.L. Vanderhart, Anal. Chem. 73, 1252-1262 (2001)

[33] R.J. Goldschmidt, S.J. Wetzel, W.R. Blair, C.M. Guttman, J. Am. Soc. Mass Spectrom. 11, 1095-1106 (2000)

[34] J. Devore, R. Peck, "Statistics: The Exploration and Analysis of Data" Belmont, California: Wadsworth Publishing Co., 1997.

[35] R. E. Kirk, "Experimental Design: Procedures for the Behavioral Sciences" Pacific Grove, California: Brooks/Cole Publishing Co., 1995.

Tables

Table 1: Source of Uncertainty for Light Scattering Measurements on SRM 2888

Contribution as Expanded Uncertainty	Uncertainty [10^3 g/mol]
Standard Deviation of Mean of Measurement	0.12
Solvent Index of Refraction	0.01
Effect of Small Molecule Impurties on dn/dc	0.13
Calibration of Refractometer	0.14
Differential Refractive Index	0.24
Wavelength of Light	0
Rayleigh Ratio of Scattering Standard, benzene	0.42
Light Polarizers	0.03
Solvent Density	0.01
Solute Masses and Solvent Masses	0.06
Reflection Correction	0.01
Instrumental Misalignment	0.01
Refraction Correction	0.04
Anisotropy of Polymer in Solution	0.01
Cutoff of Virial Expansion	0.2
Solute Degradation	0
Square root of sum of squares	0.57

Table 2: Moments of MMD from interlaboratory Comparison using MALDI-ToF-MS

M	M_n (g/mol)	M_w (g/mol)	M_z (g/mol)
Total Mean	6610	6740	6860
Standard Deviation	120	110	100
Standard Uncertainty	11.8	10.5	10.1

Figures

FT-IR Spectrum of SRM 2888

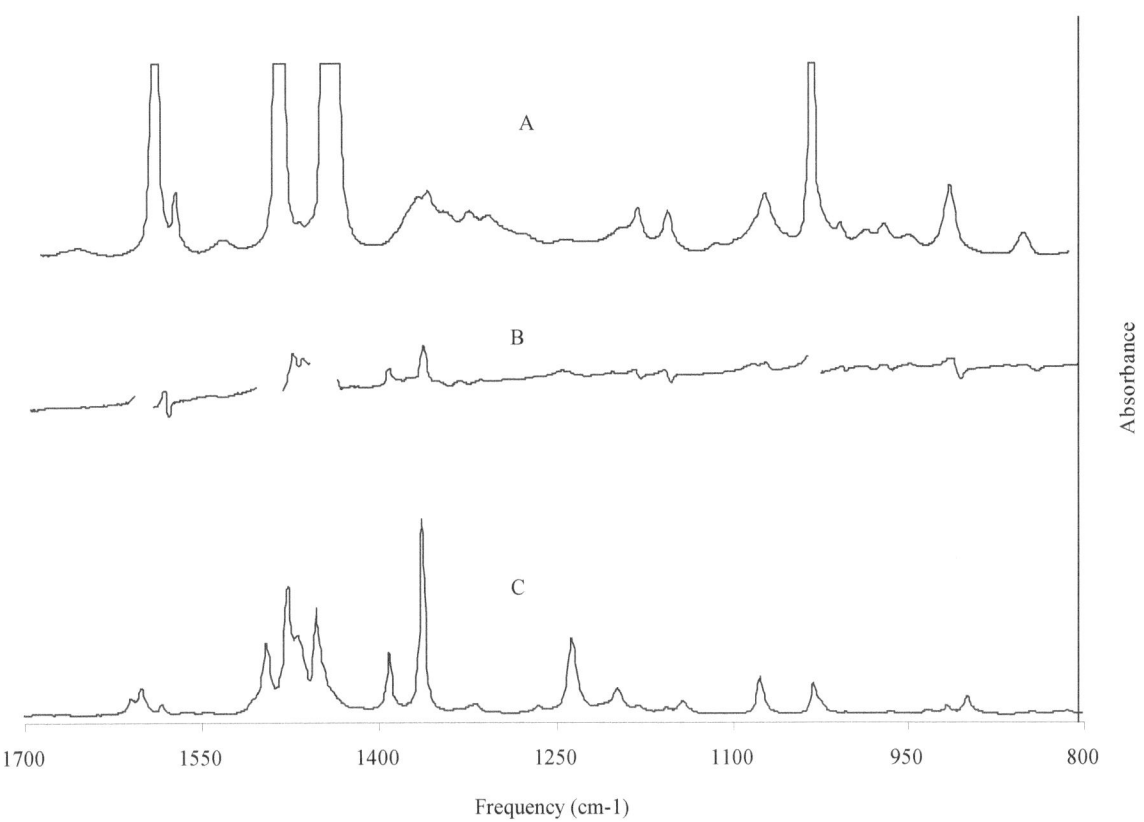

Figure 1. The FT-IR spectrum of SRM 2888 (A), neopentyl benzene (C), and the difference spectrum: SRM 2888 – SRM 1479 (B). Peaks identifying the *tert*-butyl end groups are 1365 cm^{-1}, 1393 cm^{-1}, and 1475 cm^{-1}.

Mass Distribution of SRM 2888 by Mass Spectrometry

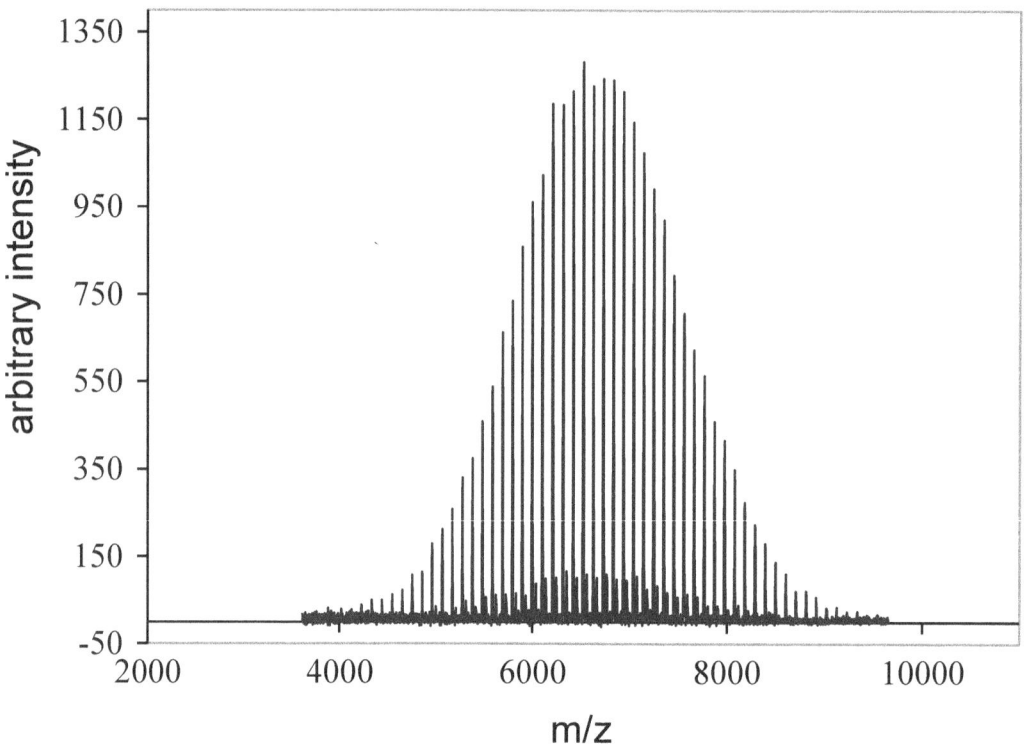

Figure 2. MALDI-TOF-MS MMD of Interlaboratory Comparison Polystyrene (SRM 2888) using the specified recipe of retinoic acid and AgTFA.

Secondary Series in Mass Spectrum of SRM 2888

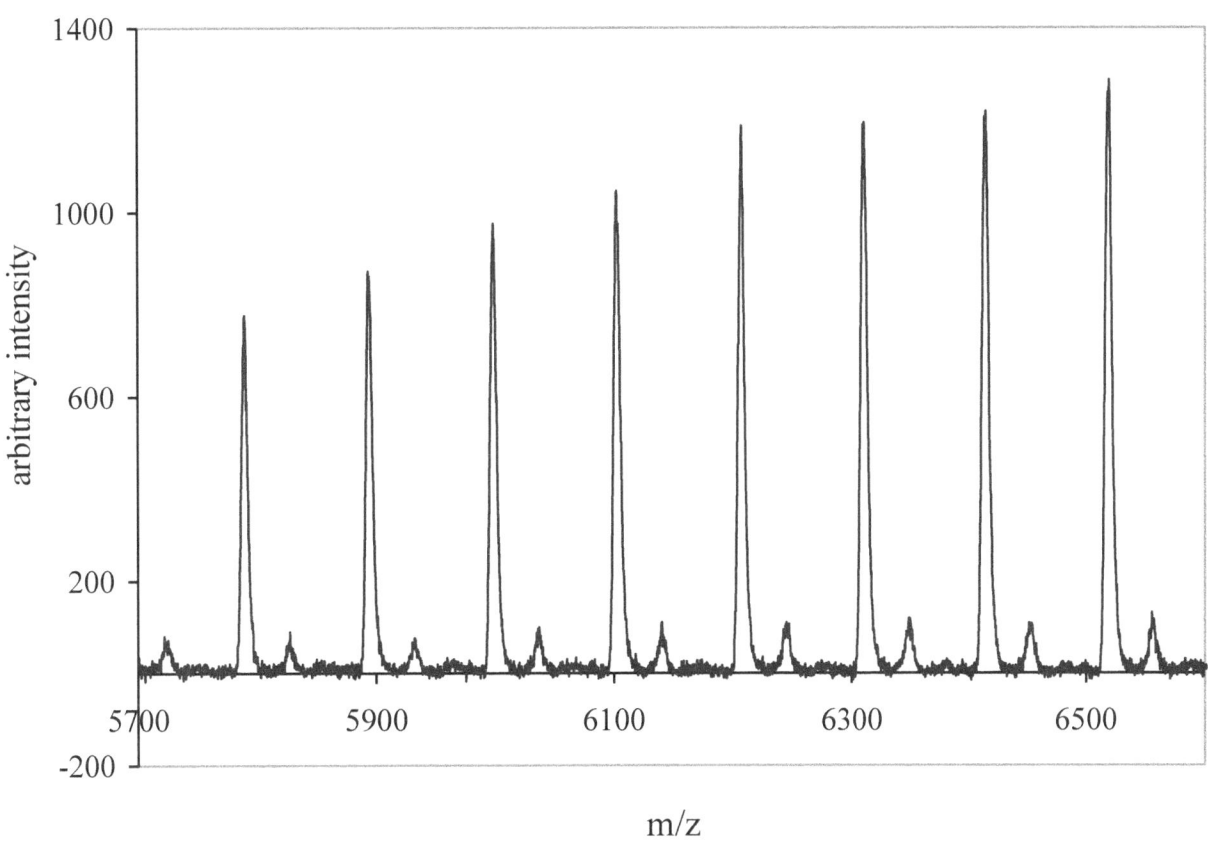

Figure 3. MALDI-TOF-MS MMD of PS expanded to show the secondary peak series.

Distribution of Mass of SRM 2888

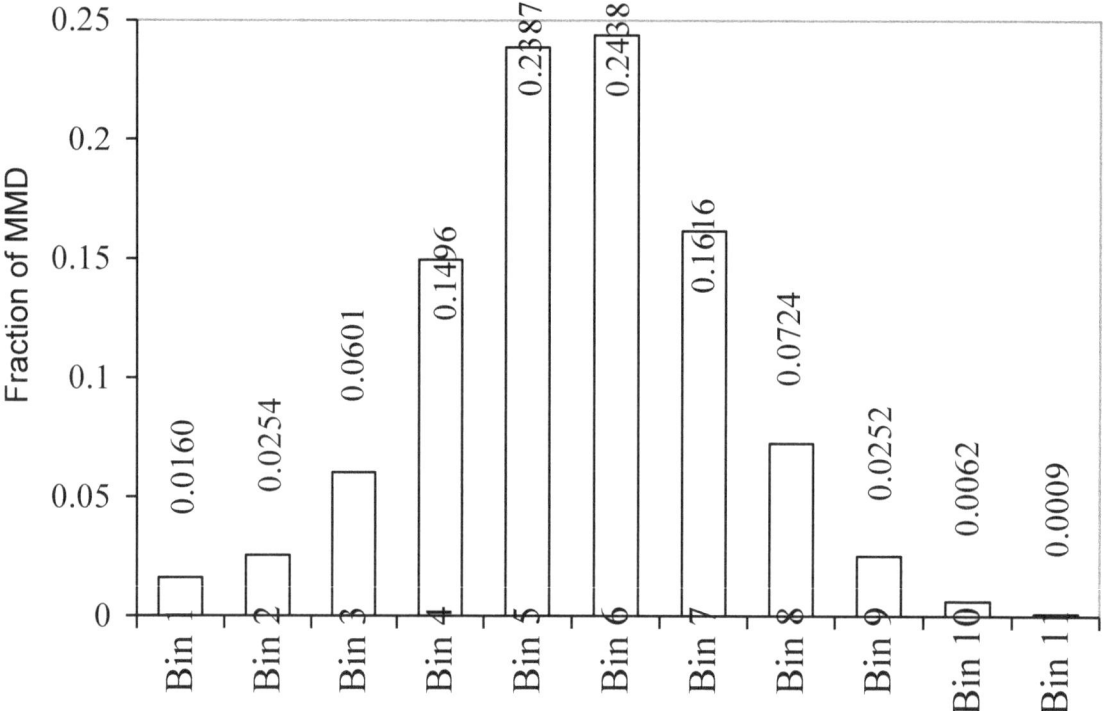

Figure 4. Histogram of the distribution of the mean bins. The fraction of the MMD is given on the histogram as well.

Distribution of M_n Reported in Interlaboratory Comparison

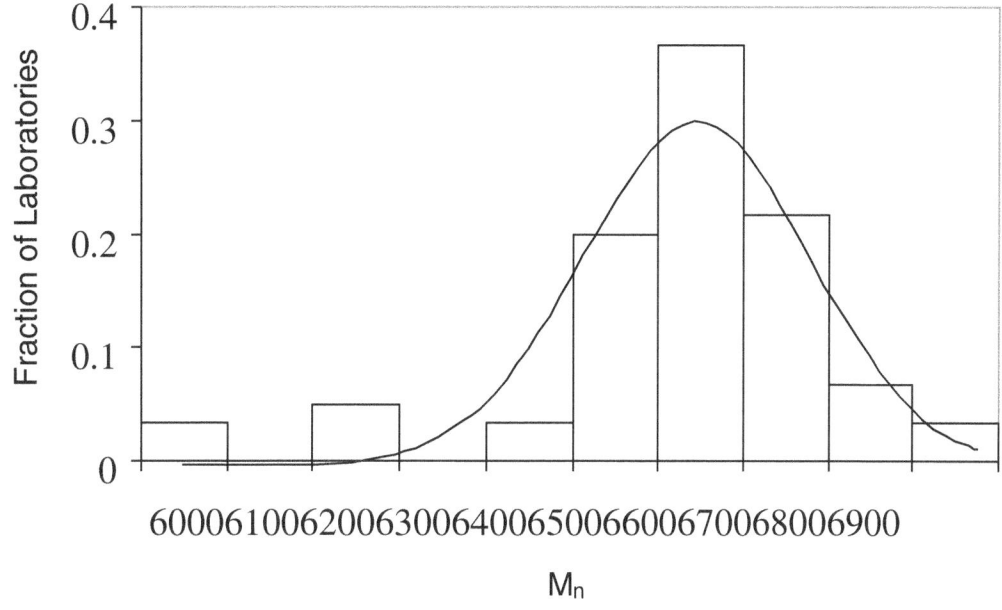

Figure 5. Histogram representing the distribution of M_n. The graphed line represents the normal distribution for the data, and the outliers are seen at 6200 u.

Distribution of End Group Mass Reported in Interlaboratory Comparison

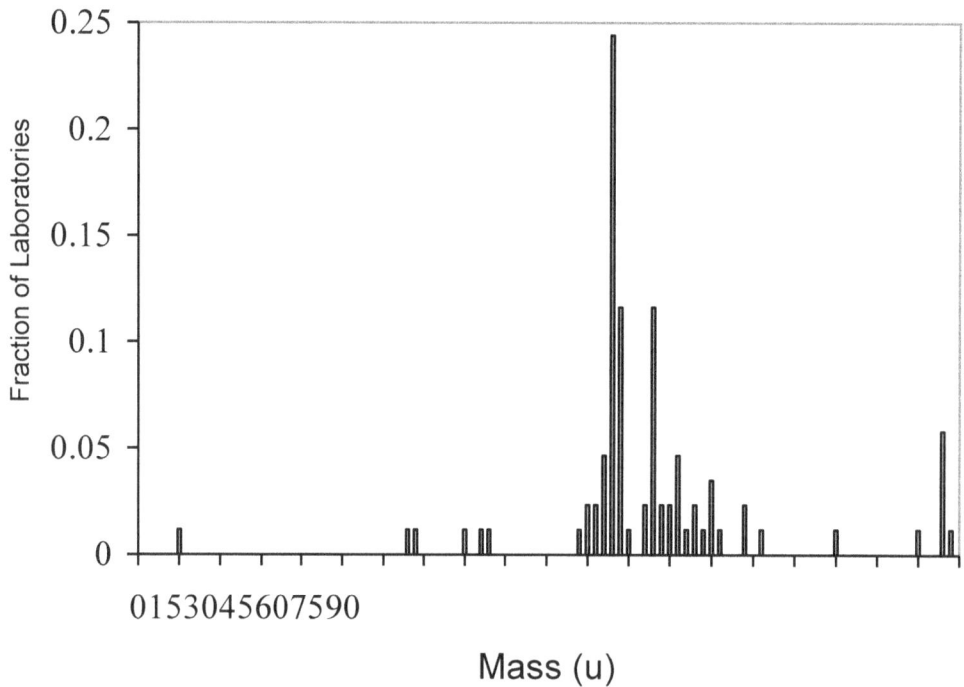

Figure 6. Distribution of the end group masses, which were calculated from the maximum peak values. The end groups, which are hydrogen and *tert*-butyl, should have a mass of 58 u.

Appendices

Appendix I: Participating Laboratories In Interlaboratory Comparison

Botten, David S.; Lab Connections Inc.
Carmen, Howard S.; Spectrometry Applications Research Laboratory
Goldschmidt, Robert; Polymers Division, NIST
Hanton, Scott D.; Air Products and Chemicals, Inc.
Haverkamp, J.; Faculty of Chemistry, Utrect University, Bijvoet Center for Biomolecular Research
Mehl, John; Hercules, David; Chemistry Department, Vanderbilt University
Jackson, Tony; Molecular Spectroscopy Team, ICI Technology
Ji, Hellen; Mays, Jimmy; Department of Chemistry, University of Alabama-Birmingham
Kowalski, Paul; Bruker Daltonics
Lattimer, Robert; BF Goodrich
Li, Liang; Department of Chemistry, University of Alberta
McCarley, Tracy, Department of Chemistry, Louisiana State University
McEwen, Charles; Dupont Corporate Center for Analytical Sciences, Central Research & Development
Montaudo, Maurizio; Department of Chemistry, CNR-Institute for Chemistry and Technology of Polymeric Materials
Powell, Brian; Analytical Sciences Group -Spectroscopy, Zeneca Specialties
Price, Phillip C.; Union Carbide Corp., Central Research and Development
Schmidt, M.; Maskos, Michael; University Mainz
Wallace, William; Polymers Division, NIST
Weidner, Steffen; BAM - Federal Institute for Material Research & Testing
Weil, David A.; 3 M Corporate Research Labs
Wetzel, Stephanie; Girard, James E.; Department of Chemistry, American University, and Polymers Division, NIST
Wilkerson, Charles W. Jr.; Analytical Chemistry, Los Alamos National Laboratory
Wu, Kuang Jan; Analytical Research, Charles Evans & Associates

Appendix II: Uncertainty in PS Low M_w due to dn/dc Varying as a Function of Molecular Mass

Following Bushuk and Benoit for the variation of light scattering from a copolymer with varying composition we obtain by their eq 5

$$R = K \sum (dn/dc)_i^2 M_i c_i \qquad (A1)$$

where $(dn/dc)_i$ is the refractive index change from molecule of mass M_i and concentration c_i. If $(dn/dc)_i$ is independent of i we get

$$R = K (dn/dc)_0^2 \sum M_i c_i = K (dn/dc)_0^2 M_w c_0 \qquad (A2)$$

where $c_0 = \sum c_i$, the normal result for light scattering. Now for varying dn/dc with end groups we assume

$$(dn/dc)_i = A + B/M_i \qquad (A3)$$

With this functional form and eq A1 for the light scattering

$$R/K = A^2 \sum M_i c_i + 2 A B \sum c_i + B^2 \sum c_i/M_i \qquad (A4)$$

$$R/K = A^2 M_w c_0 + 2 A B c_0 + B^2 \sum c_i/M_i \qquad (A5)$$

But

$$\sum c_i/M_i = c_0/M_n \qquad (A6)$$

so that

$$R/K = A^2 M_w c_0 + 2 A B c_0 + B^2 . c_0/M_n \qquad (A7)$$

This an be rewritten as

$$R/K = A^2(M_w - M_n) c_0 + M_n c_0 ((dn/dc)_{ave})^2 \qquad (A8)$$

where average dn/dc for the molecule is

$$(dn/dc)_{ave} = A + B/ M_n \qquad (A9)$$

and difference between $(dn/dc)_{ave}$ and dn/dc for large molecular mass species is $+ B/M_n$

Modifying eq (A8) one more step

$$R/K = A^2 (M_w - M_n) c_0 - (M_w - M_n)((dn/dc)_{ave})^2 c_0 + M_w ((dn/dc)_{ave})^2 c_0 \quad (A10)$$

The measured R/K gives an apparent M_{w_app}

$$R/K = M_{w\_app} \, c_0 \, ((dn/dc)_{ave})^2 \quad (A11)$$

From A11 and the above equations, we obtain

$$M_{w\text{-app}} = M_w + (M_w - M_n)(A^2/((dn/dc)_{ave})^2 - 1) \quad (A12)$$

This is the final result.

For SRM 2888:
- $A = 0.108$
- $(dn/dc)_{ave} = 0.103$
- $(M_w - M_n) = 140$ g/mol (from classical methods)
- $(M_w - M_n) = 130$ g/mol (from MALDI MS)

Taking the value 140 g/mol for $(M_w - M_n)$ the last term on the right hand side of eq. (A12) yields a value of 23 g/mol.

This is then the estimate of the uncertainty arising from using an average (dn/dc) in the light scattering equations, rather than a varying one.

References

W. Bushuk, H. Benoit, Can Jour of Chem **36** (1958) pg 1616-26; also in 'Light Scattering from Dilute Polymer Solutions", ed McIntyre and F. Gornick, Gordon and Breach (1964)

Appendix III: Estimation of (dn/dc) of a Small Amount of Cyclohexane in Toluene

Estimating the refractive index change upon mixing two organics liquids has been much studied [1,2]. Most equations work well on solutions of organic liquids. The simplest equation provided by Nelson [1] taking, as usual, $\varepsilon = n^2$

$$n_{12} = n_1 v_1 + n_2 v_2$$

where n_1 = the analyte refractive index
v_1 = the analyte volume fraction
n_2 = the solvent refractive index
v_2 = the solvent volume fraction
n_{12} = the refractive index of the mixture
and
$$v_1 = V_1 / (V_1 + V_2)$$

where V_i is the total volume of i with i =1 for analyte and 2 for solvent.

$$dn_{12}/dv_1 = (n_1 - n_2)$$

but for a dilute solution of analyte in a solvent $c_1 = V_1 \ast d_1/(V_1 + V_2) = v_1 d_1$, where d_1 is the density of the analyte.

Then $dn_{12}/dc_1 = (n_1 - n_2)/d_1$
For toluene as solvent and cyclohexane as analyte we obtain
$$(n_1 - n_2) = 1.4262 - 1.4969$$

$$d_1 = 0.7785 \text{ g/mL}$$

$$dn_{12}/dc_1 = -0.0707/0.7785 = -0.0908$$

[1] S.O. Nelson, IEEE Trans. on Elect. Insulation 26, 845-869 (1991)
[2] C.J.F. Böttcher, "Theory of Electric Polarization", Elsevier Publishing Co., 1952, Amsterdam

Appendix IV: Certificate of Analysis, SRM 2888

(See following pages.)

http://ts.nist.gov/ts/htdocs/230/232/232.htm

National Institute of Standards & Technology

Certificate of Analysis

Standard Reference Material® 2888

Polystyrene

(M_w, 7 190 g/mol)

This Standard Reference Material (SRM) is intended primarily for use in calibration and performance evaluation of instruments used to determine the molar mass and molar mass distribution by size exclusion chromatography (SEC). A unit of SRM 2888 consists of approximately 0.3 g of polystyrene powder.

Certified Value

Mass-average molar mass[*] (M_w): $7.19_x 10^3$ g/mol ± $0.57_x 10^3$ g/mol

[*]Expressed as molar mass, previously expressed as weight average molecular weight [1].

Certified Uncertainties: The certified measurement uncertainty is expressed as a combined expanded uncertainty with a coverage factor $k = 2$, calculated in accordance with NIST procedure [2]. Type A and Type B contributions to the expanded uncertainty of the measured mass-average molar mass include the uncertainties in Rayleigh ratio of the scattering standard, optical alignment, and calibration of the differential refractometer.

Expiration of Certification: The certification of SRM 2888 is valid, within the measurement uncertainties specified, until 31 January 2010, provided that the SRM is handled in accordance with the storage instructions given in this certificate. This certification is nullified if the SRM is modified or contaminated.

Maintenance of SRM Certification: NIST will monitor this SRM over the period of its certification. If substantive technical changes occur that affect the certification before expiration of this certificate, NIST will notify the purchaser. Return of the attached registration card will facilitate notification.

Technical coordination leading to certification of this SRM was provided by B.M. Fanconi of the NIST Polymers Division.

Technical measurement and data interpretation were provided by C.M. Guttman, W.R. Blair, B.M. Fanconi, R.J. Goldschmidt, W.E. Wallace, S.J. Wetzel, and D.L. Vanderhart of the NIST Polymers Division.

Statistical consultation for this SRM was provided by S.D. Leigh of the NIST Statistical Engineering Division.

The support aspects involved in the preparation, certification, and issuance of this SRM were coordinated through the Standard Reference Materials Program by B.S. MacDonald of the NIST Measurement Services Division.

Eric J. Amis, Chief
Polymers Division

Gaithersburg, MD 20899
Certificate Issue Date: 09 May 2003

John Rumble, Jr., Chief
Measurement Services Division

Storage: The SRM should be stored in the original bottle with the lid tightly closed under normal laboratory conditions.

Homogeneity and Characterization: The homogeneity of SRM 2888 was tested by SEC analysis of solutions in tetrahydrofuran at 40 °C. The further characterization of this polymer is described in reference 3.

Supplemental Information: The number-average molar mass (M_n) was determined by nuclear magnetic resonance (NMR) analysis of the end groups and found to be 6.96 $\times 10^3$ g/mol with an estimated uncertainty of 0.40 $\times 10^3$ g/mol. Fourier transform infrared spectroscopy (FTIR) and matrix-assisted laser desorption ionization time of flight mass spectrometry (MALDI-TOF-MS) were used to analyze end groups on the polymer. Only one set of end groups was found on the polymer. This polystyrene was also used in an interlaboratory comparison for the measurement of molecular mass distribution by MALDI-TOF-MS. The MALDI-TOF-MS interlaboratory comparison yielded an M_n of 6.61 $\times 10^3$ g/mol with a standard deviation of 0.12 $\times 10^3$ g/mol and M_w of 6.74 $\times 10^3$ g/mol with a standard deviation of 0.11 $\times 10^3$ g/mol. Twenty-three laboratories took part in this study [4]. A representative MALDI-TOF-MS spectrum of SRM 2888 is given in Figure 1.

NIST Certification Method: The certified value for M_w was measured on SRM 2888 using static light scattering in toluene as solvent at 23 °C [3].

REFERENCES

[1] Taylor, B.N.; *Guide for the Use of the International System of Units (SI)*; NIST Special Publication 811, 1995 Ed. (April 1995).
[2] *Guide to the Expression of Uncertainty in Measurement*; ISBN 92-67-10188-9, 1st Ed. ISO; Geneva, Switzerland, (1993); see also Taylor, B.N.; Kuyatt, C.E.; *Guidelines for Evaluating and Expressing the Uncertainty of NIST Measurement Results*; NIST Technical Note 1297, U.S. Government Printing Office. Washington, DC (1994); available at http://physics.nist.gov/Pubs/.
[3] Guttman, C.M.; Blair,W.R.; Fanconi, B.M.; Goldschmidt, R.J.; Wallace, W.E.; Wetzel, S.J.; Vanderhart, D.L.; *Certification of a Polystyrene Synthetic Polymer, SRM 2888*; NIST Special Publication 260-152 (in press).
[4] Guttman, C.M.; Wetzel, S.J.; Blair, W.R.; Fanconi, B.M.; Girard, J.E.; Goldschmidt, R.J.; Wallace, W.E.; VanderHart, D.L.; *NIST-Sponsored Interlaboratory Comparison of Polystyrene Molecular Mass Distribution Obtained by Matrix-Assisted Laser Desorption/Ionization Time-of-Flight Mass Spectrometry: Statistical Analysis*; Analytical Chemistry, Vol. 73, pp. 1252-1262 (2001).

Users of this SRM should ensure that the certificate in their possession is current. This can be accomplished by contacting the SRM Program at: telephone (301) 975-6776; fax (301) 926-4751; e-mail srminfo@nist.gov; or via the Internet http://www.nist.gov/srm.

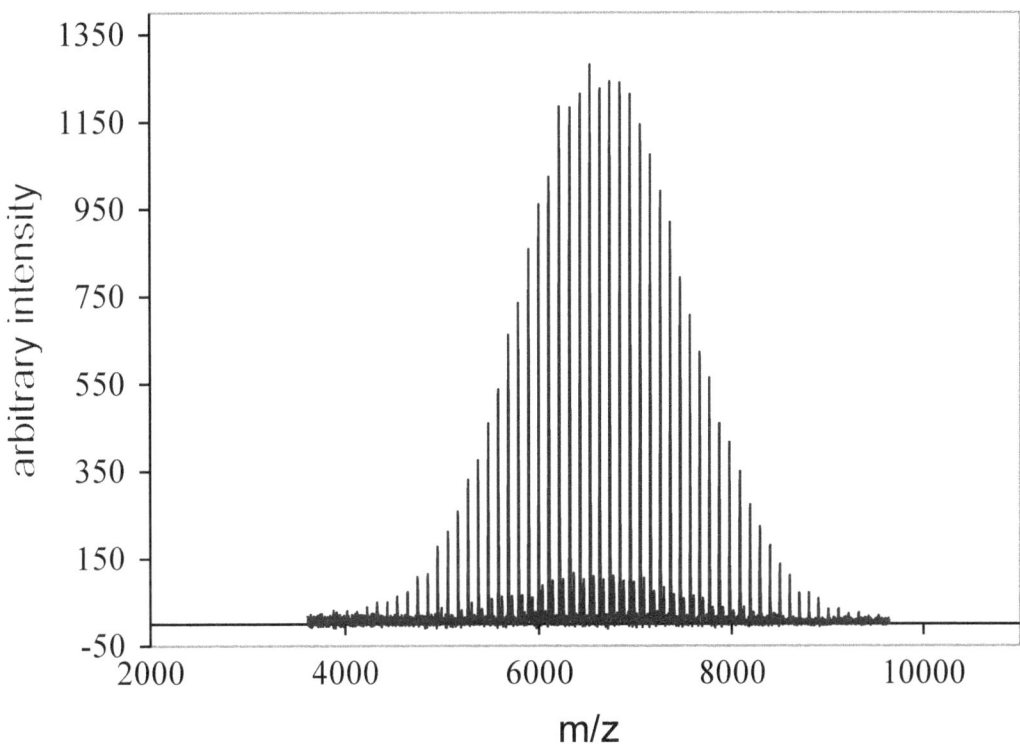

Figure 1. Mass spectrum of SRM 2888 measured by matrix assisted laser desorption/ionization mass spectrometry. Retinoic acid was used as the matrix and silver trifluoroacetate as the catonization salt.

www.ingramcontent.com/pod-product-compliance
Lightning Source LLC
Chambersburg PA
CBHW081740170526

45167CB00009B/3883